T0207267

Lecture Notes in Computer Science　　　13850

Founding Editors

Gerhard Goos
Juris Hartmanis

Editorial Board Members

The series Lecture Notes in Computer Science (LNCS), including its subseries Lecture Notes in Artificial Intelligence (LNAI) and Lecture Notes in Bioinformatics (LNBI), has established itself as a medium for the publication of new developments in computer science and information technology research, teaching, and education.

LNCS enjoys close cooperation with the computer science R & D community, the series counts many renowned academics among its volume editors and paper authors, and collaborates with prestigious societies. Its mission is to serve this international community by providing an invaluable service, mainly focused on the publication of conference and workshop proceedings and postproceedings. LNCS commenced publication in 1973.

Marina Gavrilova · C. J. Kenneth Tan ·
Mark Coates · Yaoping Hu · Henry Leung ·
Arash Mohammadi · Konstantinos N. Plataniotis ·
Helder Rodrigues de Oliveira
Editors

Transactions on Computational Science XL

Editors-in-Chief

Marina Gavrilova (iD)
University of Calgary
Calgary, AB, Canada

C. J. Kenneth Tan
Sardina Systems OÜ
Tallinn, Estonia

Guest Editors

Mark Coates
McGill University
Montreal, QC, Canada

Yaoping Hu
University of Calgary
Calgary, AB, Canada

Henry Leung
University of Calgary
Calgary, AB, Canada

Arash Mohammadi
Concordia University
Montreal, QC, Canada

Konstantinos N. Plataniotis
University of Toronto
Toronto, ON, Canada

Helder Rodrigues de Oliveira
University of Calgary
Calgary, AB, Canada

ISSN 0302-9743   ISSN 1611-3349 (electronic)
Lecture Notes in Computer Science
ISSN 1866-4733   ISSN 1866-4741 (electronic)
Transactions on Computational Science
ISBN 978-3-662-67867-1   ISBN 978-3-662-67868-8 (eBook)
https://doi.org/10.1007/978-3-662-67868-8

This Springer imprint is published by the registered company Springer-Verlag GmbH, DE,
part of Springer Nature
The registered company address is: Heidelberger Platz 3, 14197 Berlin, Germany

Editorial

The *Transactions on Computational Science* is published as part of the Springer series *Lecture Notes in Computer Science*, and is devoted to a range of computational science issues, from theoretical aspects to application-dependent studies and the validation of emerging technologies.

It focuses on original, high-quality research in the realm of computational science in parallel and distributed environments, encompassing the theoretical foundations and the applications of large-scale computations and massive data processing. Practitioners and researchers share computational techniques and solutions, identify new issues that shape the future directions for research, and enable industrial users to apply the presented techniques.

The current issue is devoted to research on Trustworthy Technologies for Autonomous Human-Machine Systems. Recently, there has been a growing interest in theoretical aspects of secure systems design, modeling, analysis, as well as deployment of human-machine systems into government, defense, and commercial applications. Researchers have successfully ventured into a wide range of information security and trustworthy computing tasks, such as information fusion, visual tracking, surveillance, biometrics, human behavioral patterns, medical image processing, human-computer interaction, robotics, and many others.

This issue, comprised of seven papers, is devoted to novel techniques for Trustworthy Technologies for Autonomous Human-Machine Systems. They include emerging and innovative applications of computer security-based applications, as well as relevant theoretical contributions.

The first article of the issue is an invited article "Challenges in Understanding Trust and Trust Modeling: Quenching the Thirst for AI Trust Management", solicited from a defense scientist Ming Hou. It outlines the newest developments and highlights rising challenges in understanding trust and in modeling trust in defense and security applications.

The second article, "Stress Contagion Protocols for Human-Autonomy Teaming", introduces a formalization of human-machine communication protocols when team members are under stress. Depending on the task developed, humans are frequently under stress. In a team, one single stressed person can pass along this state, in a contagion fashion, to the other members, causing an impact on the task outcome. To model this cascade effect, the authors have proposed a theoretical model based on the epidemiological SIR model.

The third article, "AVCA: Autonomous Virtual Cognitive Assessment", proposes an autonomous cognitive screening system based on Deep Neural Networks and Natural Language Processing, for the evaluation of patients with cognitive impairments. The model proposed evaluates hand motion to perform its evaluation. The results achieved have outperformed state-of-the-art methods.

The article "Light-Weighted CNN-Attention-Based Architecture Trained with a Hybrid Objective Function for EMG-Based Human Machine Interfaces" uses a convolution neural network approach for the development of EMG-based interfaces.

The article "Fairness, Bias and Trust in the Context of Biometric-Enabled Autonomous Decision Support" deals with a very important issue of understanding fairness, bias and trust in the development of machine-learning powered systems for decision-making support.

The article "An Autonomous Fake News Recognition System by Semantic Learning in Cognitive Computing" proposes a new methodology for fake news detection using cognitive computing. It is a profound work on a very important topic.

The final, seventh article, "Addressing Dataset Shift for Trustworthy Deep Learning Diagnostic Ultrasound Decision Support", addresses the problem of database shift, which is the impact caused by training a model with a dataset and deploying it with another one. The approach proposed is composed of several deep learning models that are selected according to the datasets. The results reported have shown higher accuracy compared to other approaches where one single method is used.

We thank all reviewers for their diligence in making recommendations and evaluating revised versions of the papers presented in this special issue. We would also like to thank all of the authors for submitting their papers to the Transactions, and the associate editors for their valuable work.

It is our hope that these articles will be a valuable resource for readers and will stimulate further research in the vibrant area of computational science theory and applications.

June 2023

Mark Coates
Yaoping Hu
Henry Leung
Arash Mohammadi
Konstantinos N. Plataniotis
Yingxu Wang
Helder C. R. Oliveira

Contents

Contents

Challenges in Understanding Trust and Trust Modeling

Quenching the Thirst for AI Trust Management

Ming Hou[✉]

Defence Research and Development Canada, Toronto, Canada
Ming.Hou@forces.gc.ca

With its broad application and profound potential, artificial intelligence (AI) technologies hold the promise of delivering transformative changes in our society. AI is becoming increasingly capable of autonomous decision-making that supports human decision makers, given human cognitive limitations including processing and monitoring large amounts of information. This AI augmented decision-making can help in areas where processing demands inhibit adequate control and understanding of complex systems or situations. However, AI is not a silver bullet for eliminating human errors. Recent catastrophic accidents in civil aviation and transportation industries are stark reminders of how over reliance on AI-driven autonomous systems can have devastating consequences [1, 2]. Embracing a philosophy that delegates human-machine system (HMS) control from the human to the machine is fraught or even dangerous with significant and complex AI technical barriers including but not limited to perceptual limitations, hidden biases [3, 4], and lack of explainability of the decision-making process [5]. The problem of trustworthy AI for autonomous HMSs is complex and multidimensional [6–8]. It is largely dependent on both effective human-machine interaction and robust and reliable AI systems. Building and maintaining trust in AI is problematic and requires solutions to encourage acceptance by both the general public and users in specialized domains.

Given its dynamic, psychosocial, transactional, and contextual nature of trust [9], a trust management system (TMS) as a viable solution is essential for any responsible AI-enabled HMSs. To build the TMS with a specific domain application, the following steps are essential as a pragmatic approach: 1) trust requirements need to be generated for the system design; 2) related trust factors influencing human trust in AI must be well understood; and 3) appropriate trust models including mathematical models must be developed for the development of the TMS.

From a systems engineering perspective, a design framework is critical to guide the design process and generate trust requirements. A concept of IMPACTS model and the related seven trust dimensions as the trustworthy human-AI teaming (HAT) requirements offers the basis of a trust design framework [6]. The IMPACTS model has been validated conceptually in a large-scale military exercise in the HAT context for managing weapon engagement processes [6, 9]. The seven characteristics of IMPACTS are:

Intention: AI systems must behave in a way that is aligned with human's intentions and ethical norms and values;

M. Gavrilova et al. (Eds.): *Transactions on Computational Science XL*, LNCS 13850, pp. 1–5, 2023.
https://doi.org/10.1007/978-3-662-67868-8_1

Measurability: AI behaviors, actions, and patterns must be observable and measurable for gauging its intentions;

Performance: AI must exhibit reliable, robust, and predictable behaviors to maximize system performance;

Adaptivity: AI must learn, understand, and adapt to situational or environmental changes, the system and task status, the human partner's mental states and performance, and guard the human resources and time to achieve the team's common goals;

Communication: AI must facilitate bi-directional communications through a Human-Machine Interface (HMI) to communicate the intentions, actions, and logic between its human partner and itself;

Transparency: AI behaviors, intentions and logic must be explained to its human partner at the right time (e.g., at an optimal workload level) with the correct format and pace (e.g., intuitively understandable means and appropriate level of details) so that the human can develop an accurate mental model of the AI's intentions and end states; and

Security: AI must ensure system safety and remain protected against accidental or deliberate attacks.

Trust guides, but does not completely determine, the human's reliance on AI systems (e.g., a robot or decision-making algorithm). Humans are more likely to rely on AI systems they trust and reject AI systems they do not trust. As such, an understanding of the trust factors that influence a human's reliance on AI is key to design responsible AI systems capable of enabling safe, reliable, and collaborative human-AI partnerships.

However, the scientific and technical literature concerning trust and trust models including conceptual models for understanding trust factors is vast and increasing [10–12], comprising many more references than one could be feasibly read in a lifetime. A quick survey of Google Scholar demonstrates the pool of over a number of million articles available for several trust related keywords and keyword combinations including "Trust" (–4,500,000), "Human" AND "Trust" (–4,490,000), "Trust" AND "Artificial Intelligence" OR "AI" (– 3,310,000), and "Trust" AND "Automation" (–765,000), etc. Being knowledgeable about such a volume of information, even though there is overlap with each of the keyword combinations and recognizing that the above reference counts are only approximately accurate, would be a challenge for even a well-read expert in the field of trust research; perhaps even beyond a state-of-the-art natural language AI chatbot like ChatGPT, if the AI is trusted enough to take on the task of summarizing the information from a couple of million documents.

Unfortunately, the scientific state of knowledge in this burgeoning trust-domain literature does not readily support constructive mathematical modeling of trust [5]. To make matters even more complicated, there does not appear to be a consensus about a scientific definition of the term "trust" or what trust implies [13–17] some of this uncertainty has prompted Wubs-Mrozewicz [18] to publish a paper attempting to define trust, where the author notes that some lay users seem to consider trust predominantly as a verb, apparently unaware that trust is also a noun. The expanding interest of trust from its colloquial use and the associated traditional roots in psychology (e.g. interpersonal, social and organizational trust) to other application domains (e.g. e-commerce) is leading to further confusion in the field [13, 15] as non-psychologists contribute even broader lay usage to the conversation about trust.

In this vein, Mayer and colleagues make a noteworthy observation:

"...the failure to clearly specify the trustor and the trustee encourages the tendency to change referents and even levels of analysis, which obfuscates the nature of the trust relationship" [13], A similar statement may be made concerning the analysis of trust factor relevance without a theoretical framework for trust within which a locus of applicability is specified for categories of factors before they are assessed for relevance to trust. The teams of Hancock et al. [19, 20] and Hoff and Bashir [21], amongst others, have conducted meta-analyses of the trust literature to determine factors that are purported to be prominent antecedents of trust, but the mechanism of how each factor affects the trust construct is not addressed, even hypothetically.

To address the lack of theoretical framework of trust factor relevance and understanding of trust mechanism, a pragmatic approach is to model aspects related to trust for the development to a TMS with the guidance of IMPACTS for HMS trust requirements. Rather than focus on trust per se, the approach considers related constructs in an attempt to stay out of the argument about definitions, yet attempt to further proposals about a hypothetical structure that defines what is colloquially called trust. The following four component models are listed in Table 1 with their definitions to form the required framework for a basic TMS: Trust Value, Trust Threshold, Trust Behavior, and Trustworthiness with the context of human-robot teaming.

Table 1. Trust components definition.

Trust Components	Definition
Trust Value	Human's level of trust of the Robot
Trust Threshold	Human's assessment of the trust requirement to rely on the Robot
Trust Behavior	Human's risk-mitigation strategy to compensate for lack of trust in the Robot
Trustworthiness	The Robot's self-assessment of its own trustworthiness (a colloquial definition of the trustworthiness construct is retained)

The trust component models are a hypothetical representation based on a limited interpretation of the scientific literature. As such, they should be expected to evolve continually as more evidence is found supporting a scientific definition of trust and its characteristics. The models underlying these components are intended to be dynamic and as generic as is feasible given the limited literature considered to date; it is anticipated that the data used in the models will be context, situation or task dependent.

This preliminary modelling approach has similarities to frameworks proposed by Mayer and colleagues [13], Lee and See [22] and some models reviewed by Lewicki and colleagues [10]. The intention is to identify how relevant trust factors from numerous reviews [17, 19–22] may contribute to each of the trust component models; however, the approach will attempt to work with generalized categories rather than specific measures (similar to Hoff & Bashir [21]), recognizing the need for a model that goes beyond a specific context.

It is often maintained that trust is a context specific psychological phenomenon; it may be that work such as the research studies trust factors identified by Hancock's team [20] or Hoff and Bashir [21] may be too task specific to be useful in a HAT trust model. Nevertheless, the descriptive models of de Visser et al. [23] as well as Hoff and Bashir [21] do provide a generic framework from which to explore the concept of trust within teams or even validate the various trust factors and frameworks in a military trust violation repair experimental context. Without a scientific program to validate trust factor relationships in constructive mathematical trust models, it likely would not be productive to start developing models of human trust or an effective TMS model.

This modeling approach attempts to build on the works of de Visser et al. [23], Mayer and colleagues [13], as well as Hoff & Bashir [21] to incorporate many of their concepts in an implementation-agnostic structure, then explore implementation in a Fuzzy Logic [24, 25] computational model. The model could be developed in a mathematical modeling software package that uses a syntax designed to be compatible with MATLAB so that they can be integrated with constructive human-robot interaction models in a modeling environment to provide an experimental simulation functionality for further human-in-the-loop empirical investigations on trust factor relevance and mechanisms. The results will provide solid foundations to develop a TMS as a viable solution to address trust issues when teaming with AI-enabled sociotechnical systems. Hopefully, this pragmatic approach provides a different way of thinking to the broad scientific and technological communities to understand trust factors and then develop a responsible, reliable, and trustworthy semi-autonomous and autonomous HMS.

Declarations. The author has no financial or proprietary interests in any material discussed in this article.

References

1. Majority Staff of the Committee on Transportation and Infrastructure: The Design, Development & Certification of the Boeing 737 Max. Technical Report. D.C. House Committee on Transportation and Infrastructure, Washington (2020)
2. Helmore, E.: Tesla behind eight-vehicle crash was in 'full self-driving' mode, says driver, https://www.theguardian.com/technology/2022/dec/22/tesla-crash-full-self-driving-mode-san-francisco. Accessed 18 4 2023
3. Angwin, J., Larson, J., Mattu, S., Kirchner, L.: Machine Bias, 1st edn., pp. 254–264. Auerbach Publications (2016)
4. Dastin, J.: Amazon Scraps Secret AI Recruiting Tool That Showed Bias Against Women. 1st edn., pp. 296–299 Auerbach Publications (2018)
5. National Academies of Sciences, Engineering, and Medicine: Human-AI Teaming: State-of-the-Art and Research Needs. Technical report. Washington, DC: The National Academies Press (2022)
6. Hou, M., Ho, G., Dunwoody, D.: IMPACTS: a trust model for human-autonomy teaming. Hum.-Intell. Syst. Integr. **3**(2), 79–97 (2021)
7. Seshia, S.A., Sadigh, D., Sastry, S.S.: Toward Verified Artificial Intelligence. Commun. ACM **65**(7), 46–55 (2022)

8. Wing, J.M.: Trustworthy AI. Commun. ACM **64**(10), 64–71 (2021)
9. Hou, M., et al.: Frontiers of brain-inspired autonomous systems: how does defense R&D drive the innovations? IEEE Syst. Man Cybernet. Mag. **8**(2), 8–20 (2022)
10. Lewicki, R.J., Tomlinson, E.C., Gillespie, N.: Models of interpersonal trust development: theoretical approaches, empirical evidence, and future directions. J. Manag. **32**(6), 991–1022 (2006)
11. Atkinson, D.J., Clark, M.H.: Autonomous agents and human interpersonal trust: can we engineer a human-machine social interface for trust? Technical Report No. SS-13–07. AAAI Press (2013)
12. Barbalet, J.: The experience of trust: it's content and basis. In: Masamichi, S. (eds.) Trust in Contemporary Society, vol. 42, pp. 11–30. Brill (2019)
13. Mayer, R.C., Davis, J.H., Schoorman, F.D.: An integrative model of organizational trust. Acad. Manag. Rev. **20**(3), 709 (1995)
14. Jøsang, A., Haller, J.: Dirichlet reputation systems. In: The Second International Conference on Availability, Reliability and Security (ARES 2007), pp. 112–119 (2007)
15. Bharadwaj, K.K., Al-Shamri, M.Y.H.: Fuzzy computational models for trust and reputation systems. Electron. Commer. Res. Appl. **8**(1), 37–47 (2009)
16. Sheridan, T.B.: Individual differences in attributes of trust in automation: measurement and application to system design. Front. Psychol. **10**(7), 1117 (2019)
17. Kaplan, A.D., Kessler, T.T., Brill, J.C., Hancock, P.A.: Trust in artificial intelligence: meta-analytic findings. Hum. Factors **65**(2), 337–359 (2021)
18. Wubs-Mrozewicz, J.: The concept of language of trust and trustworthiness: (Why) history matters. J. Trust Res. **10**(1), 91–107 (2020)
19. Hancock, P.A., Billings, D.R., Schaefer, K.E., Chen, J.Y.C., de Visser, E.J., Parasuraman, R.: A meta-analysis of factors affecting trust in human-robot interaction. Hum. Factors **53**(5), 517–527 (2011)
20. Hancock, P.A., Kessler, T.T., Kaplan, A.D., Brill, J.C., Szalma, J.L.: Evolving trust in robots: specification through sequential and comparative meta-analyses. Hum. Factors **63**(7), 1196–1229 (2021)
21. Hoff, K.A., Bashir, M.: Trust in automation: integrating empirical evidence on factors that influence trust. Hum. Factors **57**(3), 407–434 (2015)
22. Lee, J.D., See, K.A.: Trust in automation: designing for appropriate reliance. Hum. Factors **46**(1), 50–80 (2004)
23. de Visser, E.J., et al.: Towards a theory of longitudinal trust calibration in human-robot teams. Int. J. Soc. Robot. **12**(2), 459–478 (2020)
24. Zadeh, L.A.: Fuzzy sets. Inf. Control **8**(3), 338–353 (1965)
25. Cox, E.: The fuzzy systems handbook: a practitioner's guide to building and maintaining fuzzy systems. AP Professional (1994)

Stress Contagion Protocols for Human and Autonomous Robot Teams

Peter Shmerko[1(✉)], Yumi Iwashita[2], Adrian Stoica[2],
and Svetlana Yanushkevich[1]

[1] Schulich School of Engineering, University of Calgary, Calgary, AB, Canada
{peter.shmerko,syanshk}@ucalgary.ca
[2] Jet Propulsion Laboratory, California Institute of Technology, Pasadena, USA
{yumi.iwashita,adrian.stoica}@jpl.nasa.gov

Abstract. The **objective** of this paper is to formalize the stress contagion protocols in first responder teams which respond to life-threatening
crises on a daily basis. Such teams include both human and autonomous
machine teammates. While the stress contagion is an attribute of human
relations, collaboration between the humans under stress and, and the
stressed humans and machines, impacts the mission performance. Thus,
it is important to model this process and use the modeling to train the
first responder teams. This paper proposes a framework for modeling the
human-machine communication protocols using the advances in computational multivalued logic. We also separate the formalization of the protocol
from the choice of an appropriate model. This separation allows to mitigate the constraints posed by the limited real-world initial data of absence
of thereof. The proposed formalization based on multivalued logic is well-
supported by the computational and benchmarking tools.

1 Introduction

Team communication protocols aim at a specification of the interaction patterns between teammates. These protocols are the core of modeling the combat missions. Real-world combat protocols are composed of simpler protocols
that cater for some specific functionality human and autonomous machines as
teammates [12]. Since emotions (happiness, fear, madness, panic, stress) are
quintessence of human communication, intelligent robot (as teammates) must
recognize human emotions and act accordingly, e.g., decrease the trust level,
and/or increase the risk level if stress features are detected in communication
with this human.

First responders such as firefighters [15,18,51], police officers [25], law enforcement [19], tactical operators, search and rescue, and military special operations
forces [13] are responding to life-threatening crises on a daily basis [27,36,54,59].
This exposure places stress on human teammates. Combat stressors include trau-
mas, such as injury, attempted attack on one's unit or camp, killing, witnessing
death, and death of a unit member. To train the teams, their mission protocols
and potential rare events must be modeled.

M. Gavrilova et al. (Eds.): *Transactions on Computational Science XL*, LNCS 13850, pp. 6–25, 2023.
https://doi.org/10.1007/978-3-662-67868-8_2

This paper addresses the problem of stress contagion in human-robot teams, working towards a common goal, or mission. Stress may propagate between human teammates, and stressed humans may break the protocols of communication with autonomous robots. Either situation can lead to the mission failure.

Real-world data on contagion processes in first responder teams is generally unavailable. However, real-world communication protocols (in the form of a set of rules and/or regulations) are available and supply valuable information for modeling the team functioning to achieve the set goals. For example, there are regulations on communication between humans and autonomous robots [14, 24, 60], and team behavior in critical situations [10, 49].

As the level of machine intelligence increase, the machine becomes more than an automated tool and moves to the category of a teammate [8, 29, 53]. The communication between the human and robot teammates reaches a new level.

The problem of human and autonomous machines as teammates has been intensively studied from various directions, including human-machine interaction transparency [24, 43], degree of autonomy [45, 50], and trust [34, 35, 37]. In our study, we assume that the robot cannot experience stress, but it is able to recognize whether the human teammate is stressed.

This paper proposes a framework for modeling the human-machine communication protocols using the advances in computational multivalued logic. Instead of binary logic (operating with the values of 0,1), multivalued logic provides possibilities for detailed encoding stress states of teammates and decision on mission, using multiple values $(0, 1, \ldots, k)$.

We separate the formalization of the protocol from the choice of an appropriate model (Fig. 1). This is rationalized for two reasons. First, it allows to mitigate the constraints posed by the limited real-world initial data of absence of thereof. Thus, the problem is shifted to the higher level of abstraction but with well identified borders of uncertainties using the protocols. Second, the choice of modeling techniques is made simpler since the communication protocols are formalized in a realistic manner. Such separation of the protocols and models is a step towards creation of the standard benchmarks for comparison of various approaches.

This paper is organized as follows. Related works are reviewed in Sect. 2. The problem is formulated in Sect. 3. Basic definitions are given in Sects. 4. In Sect. 5, the computational guidelines are introduced. We discuss the results and provide conclusions in Sect. 6.

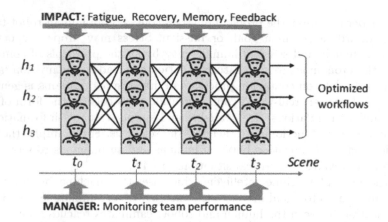

Fig. 1. Our approach to model stress contagion in first responder teams is to separate the protocol formalization (red box, focus of this paper) and the choice of the model. (Color figure online)

2   Related Work

Stress caused by a combination of events becomes a highly destructive factor that can be propagated between teammates and significantly impact the mission [55]. Integration of the autonomous machine in the team of humans require significant correction of known approaches for stress contagion, since the machines do not have emotions.

Our approach can be viewed as agent based modeling [3,14]. The novelty of our approach is in the contagion mechanism that we embed in the agent-based model using advances in computational logic.

The related works on contagion protocols are summarized Table 1. Computational epidemiological models of disease spread provides us with a standard toolkit for modifications and application. We refer to two basic contagion models [26,38]: Susceptible-Infected-Susceptible (SIS) model that uses two-valued states of a subject: susceptible (S) and infected (I); and Susceptible-Infected-Recovered (SIR) one that uses three-valued states: susceptible (S), infected (I), and Recovered (R). Further multi-valued extensions of these models are possible by adding additional factors, e.g. evacuation, immunization using demographic variability, vaccination, and others. It should be noted that these probabilistic models are used for large population (cities, urban areas, geographical locations with high population density) in which individuals are infected on average at a rate β and recover at a rate γ. For example, Markov chain is a common computational realization of SIS model. Epidemiological models do not account for spatially structured populations, or for variations in the transmission rates between individuals. A useful step towards the individual-level transmission was a combination of the above epidemiological models with the agent-based models [4,11,32,44,47]. Emerging topic of human-robot teaming is robot adaptation to human fatigue [46] and stressed behavior [48].

Stress contagion in a closed, homogeneous population can be interpreted accordingly the SIR model as follows: susceptible or ignorant (S, not stressed), spreader (I, under stress), and recovered or removed (R, stressed but cease to spread it).

3 Problem Formulation and Approach

The object of our investigation is a small groups of teammates consisting of strong cooperative humans and autonomous robots. These groups of specifically

Table 1. Sample of contagion protocols.

Protocol and Model	Description
Epidemiological contagion	
SIS and SIR contagion	In SIR protocol, individuals are distinguished by two states: as Susceptible (S) with the probability β, and Infected (I) with the probability γ, Individuals randomly go through the cycle of $S \leftrightarrow I$. The SIR protocol is an extension of a SIS by adding the Recovered, or Removed (R) class. Overview and analysis of SIS-based and SIR-based protocols and models [41]. Spatial-temporal view of the SIS and SIR models is represented in [30]. In [5], authors investigated the effects of cooperation between contagion processes based on SIS model. Details can be found in the textbook on computational epidemiology [38].
Emotion contagion	
Panic contagion	Modeling protocols of epidemiological contagion were applied to model emotion contagion, e.g. fear [7] and panic [33,57,58]. Some models are based on acceleration equation (interaction forces that change of velocity) [21] or adopted SIR protocol [33].
Stress contagion	In [52], SIR based models are adopted for study relationships between distress and mental disorders. Granger causality analysis was used in [42] for stress propagation. SIS model was used in [22] for emotion propagation in social networks.
Other societal contagion processes	
Ideas contagion	Models of contagion were adapted to describe various societal phenomena, e.g. rumors [17], ideas [2], and trust [6,63], as well as coupling stress (fear) with infection contagion, e.g. [23]. Quantum model was adapted in [64].
Rumor contagion	SIR protocol and model were adapted rumour propagation, optimization, and decision making [17].
Trust contagion	Review of trust computing proposed in [6]. Method from neuropsychology, known as spreading activation models, was used in[63].
Composite models	Multiple protocols are used in modeling processes such as rumor and crises events [1] and social response and disease outbreak [11].

trained teammates called first responders. They are aimed at execution of specific tasks such as fire fighting, emergency response, rescue operations, space missions, and military special forces. Stress is a common factor of their performance. It is a cognitive state affecting the three human systems vital for survival: vision, cognitive processing, and motor skill [16]. The state of stress can cause one, two, or all three primary systems to fail. These situations can be detected in the course of the teaming process, e.g. when teammates use visual or/and voice contact, joint actions and decision-making. While a high level of stress may lead to anxiety or absenteeism, a low level of such may also have undesirable consequences, such as lack of motivation [39].

Emotions contagion, namely, stress contagion, between human teammates is the critical property of the proposed model. Autonomous robots cannot experience stress by design, but they can detect stress in human, and change the pattern of collaboration ith humans t, thus impacting the contagion process. The above motivates the development of stress protocols for potentially possible combat scenarios and scenes. Such protocols must take into account various physiological and psychological aspects of stress aiming at development recommendations on autonomous robot design and training.

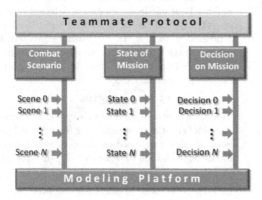

Fig. 2. Stress model: a human teammate state is initiated by stressors; the state is updated by physiological (body) and psychological (behavior) response factors of the GAS cycle.

3.1 Stress Cycle

State of stress is a dynamic phenomenon that can be modeled using a notion of perception-action cycle. Both the stressor and stress response are detectable using physiological signals (body) and psychological factors (behavior) (Fig. 2). Stress is also known as General Adaptation Syndrome (GAS) consisting of three phases [66]:

Phase I: Alarm (the immediate reaction to a stressor),

Phase II: Resistance (body adapts to the stressors), and
Phase III: Exhaustion (stress has continued for some time)

Stress is a part of human perception-action cycle (Fig. 3(a). Consider a stress monitoring assistive technology that enables a so-called support cycle. This framework contains all attributes of a cognitive dynamical system [20]. Perception-action cycle of an autonomous or semi-autonomous robot includes and is enabled by detecting the stress in human teammates (Fig. 3(b). Both GAS cycle and support cycle are performed under various risk, trust, and bias regulators, e.g. risk of stress, and trust to the autonomous machine as a teammate. Development of a control cycle o = in such as system is a challenging problem [12].

Fig. 3. Perception-action cycles of a human (*a*) and a robot teammate (*b*).

3.2 Stress Contagion in Human-Robot Teams

Given a team of four combatants (three humans h_1, h_2, and h_3, and autonomous machine m), Fig. 4 represents a potential stress contagion scenario:

- At scene t_0, team status is "Mission started". Robot m cooperates with combatant h_1.
- At scene t_1, combatant h_1 is under stressful conditions. Robot m detect the stress in h_1 and transitions to cooperation with combatant h_2. Team status is "Mission in progress".
- At scene t_2, stress propagates from h_1 to h_2. However, h_1 recovered. Robot m detects stress in h_2 and switched to cooperate with h_3 (note that choice of h_1 is possible but was not selected). Team status is "Mission in progress".
- At scene t_3, stress is propagated from 2nd h_2 to 3rd h_3 combatant, and h_2 is still in a stressed state. Robot m detects stress in both h_2 and h_3 (the majority of the humans is under stress). Team status is "Mission failed".

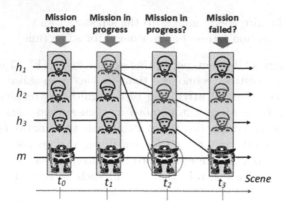

Fig. 4. Example of stress contagion scenario in human-robot team as a set of scenes.

3.3 Denotations

Given a human-machine team of first responders, the combat model consists of the following components (Fig. 5):

- A combat scenario as a set of N scenes $t_i, i = 0, 1, \ldots N$:

$$\texttt{Combat Scenario} = \{t_0, t_1, \ldots, t_N\}$$

- A state of the mission as a function ψ of N states and the corresponding scenes $s_i(t_i), i = 0, 1, \ldots N$:

$$\texttt{State of Mission} = \psi(s_0(t_0), s_1(t_1) \ldots, s_N(t_N))$$

- A decision on the mission as a function Ψ of N decisions for the corresponding states $\psi_i, i = 0, 1, \ldots N$:

$$\texttt{Decision on Mission} = \Psi(\psi_0, \psi_1, \ldots, \psi_N)$$

- A modeling platform including the methodology and the respective computational techniques.

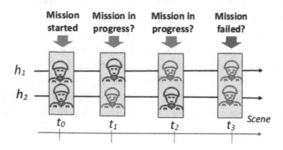

Fig. 5. Teammate protocol conveys information about the combat scenario, state of mission, and decision on mission which are placed on an appropriate modeling platform.

3.4 An Approach

The stress contagion protocol, illustrated above, is a small size network where each node corresponds to a teammate. The links between the nodes reflect relationships between the teammates. Those links are varied through the mission as well as the states of the teammates. Thus, the network structure is not stable and can evolve to achieve the specified mission. The evolution is represented inn the proposed model using the scenes, corresponding to the states. The state of stress and its contagion are considered in this model to be random processes. We make this assumption, while acknowledging that they are noit random, in general, and are defined by both the environmental conditions, psychological portrait of the combatants, team dynamics, training and other factors.

Computational logic provides a wide spectrum techniques for representation, processing, and propagation of states of a system, while the states are certain functions, where the function and its variables' values encoded using integers [61]. Computational logic is a suitable form of representation of such states, as it operates with multi-valued variables and their functions; the latter can represent certain protocols or rules of operation on variables. The core data structure to represent logic functions is the truth-table. The size of a truth table for a binary-valued function of n binary variables is 2^n. For the m-valued variables and function, this size is m^n. The truth table is useful for manipulation with logical data but it is impractical for large number of variables. The breakthrough solution for a practical representation of logic functions was proposed in 80th in the form of a graph-based data structure called decision diagrams (DDs). There are many guides on the DD construction and application, e.g. [62]. We will use this data structure in the proposed model.

4 Basic Definitions

We adapted the definition of a scenario and a scene for the description of an automated system from [56] to our purposes.

Definition 1. A combat scenario describes the temporal development between several scenes in a sequence of **combat scenes**.

Every combat scenario starts with an initial scene. Actions and events, as well as goals and values, characterize temporal development of the scenario that spans a certain amount of time.

Definition 2. Given a teammate $h_i, i = 1, 2, \ldots, n$, their state is either operational or a state of stress, encoded by 0 and 1, respectively. A **tabulated state** of teammate scenario is defined as $h_i = \sigma, \sigma_i \in \{0, 1\}$. Given n teammates $h_1, h_2, \ldots, h_n$, their tabulated states, also called scenes, are defined as $h_1 h_2 \ldots h_n = \sigma_1 \sigma_2 \ldots \sigma_n$.

Given a team of two combatants h_1 and h_2, Fig. 6 represents a potential stress scenario of four scenes:

– In scene t_0, both combatants are operational. Team status is "Mission started".
– In scene t_1, combatant h_2 is under stress. Team status is "Mission in progress".
– In scene t_2, combatant h_1 is under stress. Team status is "Mission in progress".
– In scene t_3, both combatants are under stress. Team status is "Mission failed".

An example of the tabulated states of the two teammates h_1, h_2 is given in Table 2.

Definition 3. Stress respond function $g(m) \in \{0,1\}$ of the autonomous machine m is encoded by "1" (detection of stress in a human teammate) or "0" (no stress detected).

Definition 4. Consider a team of $i + j$ teammates: humans $h_i, i = 1, 2, \ldots, n$ and autonomous machines $m_j, j = 1, 2, \ldots, k$. **State of the combat unit** is a function $f(h_i, g(m_j))$, where $g(m_j)$ is a response of a machine m_j once the stress in a human teammate cooperating with this machine is detected. This function evolves along with the human teammate state.

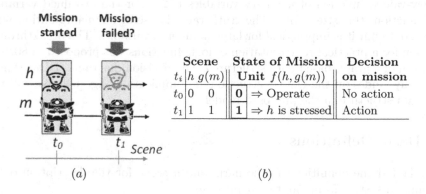

Fig. 6. Representation of a stress scenario given the two combatants and four scenes.

Table 2. Stress protocol of a combat unit consisting of two combatants h_1 and h_2.

Teammate			State of Combat Unit		Decision on mission
scene				$f(h_1, h_2)$	
t_i	h_1	h_2			
t_0	0	0	**0**	$\Rightarrow$ Regular mode	No action
t_1	0	1	**1**	$\Rightarrow h_2$ is stressed	No action
t_2	1	0	**1**	$\Rightarrow h_1$ is stressed	No action
t_3	1	1	**X**	$\Rightarrow$ Uncertain	Uncertain

Definition 5. Decision table or a combat team mission is a composite data structure that conveys three types of information:

1. Tabulated states of each teammate;
2. Tabulated states of the combat unit; and
3. A decision on the mission.

An example of the decision table is given in Table 2. Consider two combatants, $h_1, h_2 \in \{0, 1\}$, where unstressed and stressed combatant are encoded as 0 and 1, respectively. Possible states of teammate h_1 and h_2 are tabulated, they produce the $2^2 = 4$ state space. State of combat unit is represented by tabulated function $f(h_1, h_2) = [0111]^T = h_1 \vee h_2$ and a comment. The actions supporting the mission may be missed, delayed, or carried out incorrectly if a teammate's affective state is specified as stressed.

Figure 7(a) represents a potential stress scenario of two scenes:

- In scene t_0, both teammates operate in a regular mode. Team status is "Mission started".
- In scene t_1, human-teammate is over stress conditions. Team status is "Mission failed".

Stress contagion protocol for a scenario involving a human h and a machine m is given in Fig. 7(b). Machine m manifests $g(m) = 1$ that action is required if stressed human teammate h is identified. Formally, a state of a combat unit is described by the function $f(h, g(m)) = h \vee g(m)$. The unused assignments $\{01, 10\}$ are called don't cares.

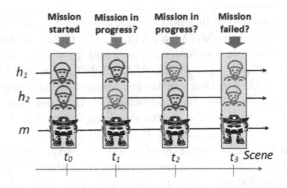

Fig. 7. Human and machine teammate scenario: (a) Illustration of a scenario consisting of two scenes and (b) Protocol formalization using a decision table.

Note that the tabulated form of logic functions is good for illustrative purposes but impractical because its size growth exponentially with the number of variables. This is the reason to use scalable graphical data structures [62].

4.1   Introduction to Multivalued Logic

An n-variable m-valued function f is defined as a mapping of a finite set $\{0,\ldots,m-1\}^n$ into a finite set $\{0,\ldots,m-1\}$, i.e., $f : \{0,\ldots,m-1\}^n \to \{0,\ldots,m-1\}$. For example, function $f : \{0,1,2\}^n \to \{0,1,2\}$ is called a ternary logic function, function $f : \{0,1,2,3\}^n \to \{0,1,2,3\}$ is called a quaternary logic function. Multivalued algebra is a generalization of Boolean algebra towards a set of m elements $M = \{0,1,2,\ldots,m\}$ and corresponding operations. A multivalued logic function f of n variables $x_1, x_2, \ldots, x_n$ is a logic function defined on the set $M = \{0,1,\ldots,m-1\}$. The logic operations defined on multivalued variables, are similar to Boolean algebra, with addition of a Literal. In particular,

$$\underbrace{\text{MAX}(x_1,x_2)}_{\text{Denoted by “+”}} = \begin{cases} x_1 & \text{if } x_1 \geq x_2 \\ x_2 & \text{otherwise} \end{cases} \qquad \underbrace{\text{MIN}(x_1,x_2)}_{\text{Denoted by “×”}} = \begin{cases} x_2 & \text{if } x_1 \geq x_2 \\ x_1 & \text{otherwise} \end{cases}$$

(a) (b)

When $m = 2$, MAX and MIN operations turn into Boolean OR ($\vee$) and AND ($\wedge$), respectively. The Literal operation is specified below

$$x^y = \begin{cases} m - 1 & \text{if } x = y, \\ 0 & \text{otherwise.} \end{cases}$$

For example, a 3-valued logic function $g = 2 \times x_1^0 \times x_2^2 + 1$ is calculated as follows: $g = MAX(2 \times x_1^0 \times x_2^2, 1) = MAX(MIN(2, x_1^0, x_2^2), 1)$. Given the assignment of the function's variables, in particular, $x_1 = 0$ and $x_2 = 0$, the function value is $g = MAX(MIN(2, 0^0, 0^2), 1) = MAX(MIN(2,2,0), 1) = MAX(0,1) = 1$.

A multivalued logic function can be represented using a truth table which tabulates all possible variables' combinations with their associated function values. A truth column vector of a multivalued logic function f of n m-valued variables $x_1, x_2, \ldots, x_n$ is defined as $\mathbf{F} = [f(0), f(1), \ldots, f(m^n - 1)]^T$. For example, the truth column vector $\mathbf{F}$ of a ternary MAX function of two variables is $\mathbf{F} = [012112222]^T$ and can be represented in algebraic form as follows: $MAX(x_1,x_2) = 0 \times x_1^0 x_2^0 + 1 \times x_1^0 \times x_2^1 + 2 \times x_1^0 \times x_2^2 + 1 \times x_1^1 \times x_2^0 + 1 \times x_1^1 \times x_2^1 + 2 \times x_1^1 \times x_2^2 + 2 \times x_1^2 \times x_2^0 + 2 \times x_1^2 \times x_2^1 + 2 \times x_1^2 \times x_2^2$. The readers can refer [62] for details on multivalued logic.

5   Formalization of the Stress Contagion Protocol

The attributes of the proposed stress contagion protocols include

1. a scenario,

2. the scenes,
3. a stress respond function,
4. a state of the combat unit, and
5. a decision table.

The dynamics of combat are reflected by these five attributes and explained using Boolean logic that operates with values 0 and 1. Binary-valued protocols are useful for preliminary exploration modeling tasks: they are simple in illustration and understanding, and well computationally supported by decision diagrams and benchmarks.

Definition 6. A stress contagion protocol for a team of first responders is a working set of rules (logical, mathematical, computational) for preparation, processing, communication, and decisions on the mission scenarios. Stress contagion protocols are embedded into appropriate **model** of stress contagion (e.g. SIS, SIR, agent based) in order to simulate the teammate performance.

The stress contagion protocols reflect best practices of first responders. The stress contagion protocols is a part of standardization – a guideline to be followed when different approaches and models are compared.

5.1   Compatibility Criteria

Binary-valued stress protocols are compatible with SIS contagion model, which is binary by definition. This compatibility provides a simple embedding of the proposed protocol to any SIS based model, e.g. rumor contagion [17]. However, the binary nature of the stress protocols is a simplification of the real-world stress contagion mechanisms. The SIR epidemiological model better fits this demand, but is generally applied to the large population with 3–5 control factors. The targeted in our study population is small, e.g. team of about a dozen of trained individuals and specifically designed autonomous robots. Their experiences with stress are different, and their communication and actions are defined. This is a motivation to use a multivalued logic to describe these protocols (Table 3).

Table 3. Qualitative comparison of contagion protocols based on Boolean logic and multiple logic.

Beneficial features	Limitations
Stress contagion protocols based on Boolean logic	
– Compatible with epidemiological binary SIS models; – Standard software packages are available for generation of the protocols; – Standard benchmarks can be adapted for comparison of the protocols	– Limited to small groups; – Limited number of functions, 2^n
Stress contagion protocols based on multivalued logic	
– Compatible with the epidemiological multi-state SIR models; – Standard software packages are available; – Standard benchmarks can be adopted for comparison.	– Limited to small groups; – Unconventional form of function description

5.2 Human Team Scenario

Let us denote $h_i = 0$ if $h_i \equiv S$; $h_i = 1$ if $h_i \equiv R$; and $h_i = 2$ if $h_i \equiv I$. Stress contagion protocol for a mission under stress contagion is given in Table 4. The state of such a combat unit is described by a 3-valued logic function as follows:

$$\text{State} = f_0 \times h_1^0 h_2^0 + f_1 \times h_1^0 h_2^1 + f_2 \times h_1^0 h_2^2 + f_3 \times h_1^1 h_2^0 + f_4 \times x_1^1 h_2^1 +$$
$$\times f_5 \times h_1^1 h_2^2 + f_6 \times h_1^2 h_2^0 + f_7 \times h_1^2 h_2^1 + f_8 \times h_1^2 h_2^2$$
$$= h_1^0 h_2^1 + h_1^1 h_2^0 + h_1^1 h_2^1 + h_1^1 h_2^2 + h_1^2 h_2^1$$

When a susceptible individual enters into contact with an infected individual, there is a certain probability that the susceptible individual becomes infected and moves from the susceptible class into infected class. After time elapsed, every infected individual transitions from the infected state into the recovery state.

5.3 Human-Machine Team Scenario

Consider a model for a team in which the i-th human combatant h_i is in one of the following states, $h_i(t) \in 0, 1, 2$, where S, I, and R are encoded by 0,1, and 2, respectively. Assume that the team consist of two human combatants and an autonomous machine. Let 0 and 3 encode the "Operational" and "Failed" mode, respectively. The states of combatants, i.e. human and autonomous machines, as well as the state of the team mission across the scenes, are also encoded. Stress

Table 4. Stress contagion protocol for a mission based on the SIR model for 3^2 space: (a) State of two combatants h_1 and h_2 described by 3-valued functions; decision on stress contagion is made using 3-valued logic.

Scene			State of Combat Unit		Decision on mission
t_i	h_1	h_2		$f(h_1, h_2)$	
t_0	0	0	$f_0 = 0$	Operational	No action
t_1	0	1	$f_1 = 1$	h_2 is stressed	No action
t_2	0	2	$f_2 = 0$	h_2 recovered from stress	No action
t_3	1	0	$f_3 = 1$	h_1 is stressed	Action
t_4	1	1	$f_4 = 1$	Both h_1 and h_2 are stressed	Out of operation
t_5	1	2	$f_5 = 1$	h_1 is stressed and h_2 recovered	Action
t_6	2	0	$f_6 = 0$	h_1 recovered from stress	No action
t_7	2	1	$f_7 = 1$	h_1 recovered and h_2 is stressed	Action
t_8	2	2	$f_8 = 0$	Both h_1 and h_2 recovered	No action

contagion protocol for the scenes within this SIR model is shown in Table 5.

$$g_1(h_i) = \begin{cases} 0 & \text{if } h_i \equiv S; \\ 1 & \text{if } h_i \equiv R; \\ 2 & \text{if } h_i \equiv I. \end{cases} \quad g_2(m_i) = \begin{cases} 0 & \text{if } m_i \text{ tested}; \\ 1 & \text{if } m_i \text{ operate}; \\ 2 & \text{if } m_i \text{ failed}. \end{cases} \quad f = \begin{cases} 0 & \text{if pause}; \\ 1 & \text{if continue}; \\ 2 & \text{if failed}. \end{cases}$$

Let us choose Scenario 1 of mission, $f = g_1(h_1, h_2) \odot g_2(m)$. The decision table of the term $g_1(h_1, h_2)$ is given in upper plane of Table 5. It is assumed that mission can be continued if one of human combatant is stressed. Formally, $g_1(h_1, h_2) = 1 \times (h_1^0 h_2^0 + h_1^0 h_2^1 + h_1^1 h_2^0 + h_1^1 h_2^1) + 2 \times h_1^2 h_2^2$. Table 5 represents stress contagion protocol for both human $g_1(g_1, h_2)$ and machine $g_2(m)$. The equation that describes this combat scenario is $f = 1 \times g_1^1 g_2^1 + 2 \times (g_1^2 g_2^0 + g_1^2 g_2^1 + g_1^2 g_2^2)$, where $g_1 = g_1(h_1, h_2)$ and $g_2 = g_2(m)$ (Fig. 8).

5.4   Scenario Formalization

Each combat scenario is described using variables which describe environmental information, tactic and strategy. Our work contributes to the formalization of the stress contagion as well as the mission scenario as shown below. Given two human combatants h_1 and h_2 and an autonomous machine m, the following mission scenarios $f_i, i = 1, 2, 3$, are possible:

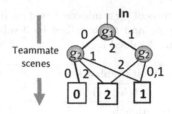

Fig. 8. Illustration of a scenario consisting of two humans, one machine, and four scenes.

Table 5. Stress contagion protocol for the mission $f = g_1(h_1, h_2) \times g_2(m)$ of two human combatants and a machine: the upper plane of decision table represent human combatants (term $g_1(h_1, h_2)$), and the lower plane of decision table combines the states of the humans and the machine.

	Scene			State of Combat Unit	Decision on mission
t_i	$g_1(h_1, h_2)$	$g_2(m)$		$f(g_1(h_1,h_2), g_2(m))$	
t_0	0	0	$f_0 = 0$	Operate	No action
t_1	0	1	$f_1 = 0$	Operate	No action
t_2	0	2	$f_2 = 0$	Machine M failed	No action
t_3	1	0	$f_3 = 0$	Human recovered	No action
t_4	1	1	$f_4 = 1$	Human recovered	No action
t_5	1	2	$f_5 = 0$	Human recovered, machine failed	No action
t_6	2	0	$f_6 = 2$	Human is stressed	Action
t_7	2	1	$f_7 = 2$	Human is stressed	Action
t_8	2	2	$f_8 = 2$	Human is stressed, machine failed	Action

$$\text{Scenario 1:} \quad f_1 = g(h_1, h_2) \odot m$$
$$\text{Scenario 2:} \quad f_2 = (h_1 \odot m) * (h_2 \odot m)$$
$$\text{Scenario 3:} \quad f_3 = (h_1 * m) \odot (h_2 * m)$$

where $\odot$ and $*$ denote the decision-making operations, and $g()$ is a function of cooperation. Scenario 1 represents the teaming with humans and machine operating separately, e.g. $g(h_1, h_2) + m$. In Scenario 2, one human is teaming up with a machine, e.g. $(h_1 \times m) + (h_2 \times m)$. Scenario 3 corresponds to cooperation of both humans with the machine, e.g. $(h_1 + m) \times (h_2 + m)$.

5.5   Software Packages and Benchmarks

The proposed stress contagion protocols are based on computational logic. The software packages and benchmarks for logic function computing, manipulation and optimization, are publicly available. Examples of such are provided below:

1. The DD packages is available from https://github.com/johnyf/tool/lists/ blob/main/bdd.md

2. Benchmarks are available from https://github.com/lsils/benchmarks.

While this work does now include the results of scenario modeling on mentioned benchmarks, it provides recommendations on the composite usage of these tools. The advanced modeling of the above scenarios can be conducted using message-passing, factor graph, Bayesian Network and related libraries are available from https://pgmpy.org/models/factorgraph.html. In particular, the Bayesian network package is available from https://pyagrum.readthedocs.io/en/1.3.2/.

Note that the development of performance measures, metrics, and benchmarks for human-machine teaming protocols is a challenging area. Review [9] identifies the demand in assessing risk and bias, human and machine states, attention, cohesion, intervention, etc. Thus, our work contributes to developing the computationally affordable and benchmarking protocols for the above tasks.

6 Summary, Conclusions, and Future Work

This work is **motivated by the demand** to study the performance of small groups of first responders, who are highly cooperative teammates, both human and robots, under condition of the contagious stress propagated among humans. These real-world stress contagion situations should be modeled in order to provide recommendations for training and predict mission failures and teammate burnouts.

There is a well-identified gap between the real-world first responder scenarios and their models. This gap is caused by the limited or absence statistics, and is addressed so far by applying the regulation rules called protocols. Our **key assumption** is that these protocols are the links between the real-world scenarios and their models. In other words, we propose to separate the protocols from the modeling paradigms. We found that this separation is beneficial for purposes of relatively small and highly connected groups operating under spatio-temporal constraints. In large population contagion models, e.g. SIS-based and SIR- based models, the protocols are rather defined for average individual (that represent a large population) with the specified contagion mechanism (model). These protocols reflect real-world statistics in a very simplified manner (e.g., the probabilities of two states in the SIS-based models and of 3–4 states in the SIR-based models).

There no "average" teammate in small first responder teams. The role of each teammate (human or robot) in each scenario and scene of the contagion process is unique. Their individual and joint contribution in stress contagion process should be clearly specified and embedded in the model. This is the core of our approach, in which we formalized such stress contagion protocols for further modeling.

The key conclusions from our work are as follows:

1. The proposed protocols can be embedded to various information exchange (contagion) process and models, i.e. social phenomena (e.g. rumors, trust, panic, fear, ideas) and contagion of biological infection.

2. Using spatio-temporal conditions, the proposed protocol can be used for modeling of various phenomena of large population, e.g. in both clustered community [30], and in the closed community [28].

Future work includes extending the protocol description by using fuzzy logic, e.g. [31] and probabilistic approaches [28,40].

References

1. Agarwal, P., Al Aziz, R., Zhuang, J.: Interplay of rumor propagation and clarification on social media during crisis events - a game-theoretic approach. Europ. J. Oper. Res. **298**, 714–733 (2022)
2. Bettencourt, L.M.A., Cintron-Arias, A., Kaiser, D.I., Castillo-Chavez, C.: The power of a good idea: quantitative modeling of the spread of ideas from epidemiological models. Phys. A **364**, 513–536 (2006)
3. Bonabeau, E.: Agent-based modeling: methods and techniques for simulating human systems. Proc. Nat. Acad. Sci. **99**(suppl. 3), 7280–7287 (2002)
4. Bosse, T., Duell, R., Memon, Z.A., Treur, J., van der Wal, C.N.: Agent-based modeling of emotion contagion in groups. Cogn. Comput. **7**, 111–136 (2015)
5. Chen, L., Ghanbarnejad, F., Brockmann, D.: Fundamental properties of cooperative contagion processes. New J. Phys. **19**, 103041 (2017)
6. Cho, J.-H., Chan, K., Adali, S.: A survey on trust modeling. ACM Comp. Surv. **48**(2) (2015). Article 28
7. Cornes, F.E., Frank, G.A., Dorso, C.O.: Fear propagation and the evacuation dynamics. Simul. Model. Pract. Theory **95**, 112–133 (2019)
8. Crandall, J., Jacob, W., Oudah, M., et al.: Cooperating with machines. Nat. Commun. **9**(1), 233 (2018)
9. Damacharla, P., Javaid, A.Y., Gallimore, J.J., Devabhaktuni, V.K.: Common metrics to benchmark human-machine teams (HMT): a review. IEEE Access **6**, 38637–38655 (2018)
10. Dietz, A.S., Driskell, J.E., Sierra, M.J., Sallie, J., Weaver, T.D., Eduardo, S.: Teamwork under stress. In: Salas, E., et al. (eds.) The Wiley Blackwell Handbook of the Psychology of Team Working and Collaborative Processes, pp. 297–315. Wiley (2017)
11. Fast, S.M., Gonzalez, M.C., Wilson, J.M., Markuzon, N.: Modelling the propagation of social response during a disease outbreak. J. R. Soc. Interface **12**, 20141105 (2015)
12. Feng, L., Wiltsche, C., Humphrey, L., Topcu, U.: Synthesis of human-in-the-loop control protocols for autonomous systems. IEEE Trans. Autom. Sci. Eng. **13**(2), 450–462 (2016)
13. Gamble, K.R., Vettel, J.M., Patton, D.J., et al.: Different profiles of decision making and physiology under varying levels of stress in trained military personnel. Int. J. Psychophysiol. **131**, 73–80 (2018)
14. Gianchetti, R., Marcelli, V., Cifuentes, J., Rojas, J.: An agent-based simulation model of human-robot team performance in military environments. Syst. Eng. **16** (2013). https://doi.org/10.1002/sys.21216
15. Grant, C., Hamins, A.P., Bryner, N.P., Jones, A.W., Koepke, G.H.: Research roadmap for smart firefighting. Nat. Inst. Stand. Technol. (NIST), Spec. Publ. **1191** (2015)

16. Grossman, D., Siddle, B.K.: Psychological effects of combat. In: Kurtz, L.R. (ed.) Encyclopedia of Violence, Peace and Conflict, 2nd ed. Elsevier (2008)
17. Gürbüz, B., Mawengkang, H., Husein, H., Weber, G.-W.: Rumour propagation: an operational research approach by computational and information theory. Central Eur. J. Oper. Res. **30**(1), 345–365 (2022)
18. Hancock, P.A.: Specifying and mitigating thermal stress effects on cognition during personal protective equipment use. Hum. Factors **62**(5), 697–703 (2020)
19. Harris, K.R., Eccles, D.W., Freeman, C., Ward, P.: 'Gun! Gun! Gun!': an exploration of law enforcement officers' decision-making and coping under stress during actual events. Ergonomics **60**(8), 1112–1122 (2017)
20. Haykin, S.: Cognitive Dynamic Systems (Perception-Action Cycle, Radar, and Radio). Cambridge University Press, New York (2012)
21. Helbing, D., Farkas, I., Vicsek, T.: Simulating dynamical features of escape panic. Nature **407**, 487–490 (2000)
22. Hill, A.L., Rand, D.G., Nowak, M.A., Christakis, N.A.: Emotions as infectious diseases in a large social network: the SISa model. Proc. Roy. Soc. B **277**, 3827–3835 (2010)
23. Jain, K., Bhatnagar, V., Prasad, S., Kaur, S.: Coupling fear and contagion for modeling epidemic dynamics. IEEE Trans. Network Scie. and Eng. **10**, 1–14 (2022)
24. Johnson, M., Bradshaw, J.M., Hoffman, R.R., Feltovich, P.J., Woods, D.D.: Seven cardinal virtues of human-machine teamwork: examples from the DARPA robotic challenge. IEEE Intel., 74–80 (2014)
25. Kelley, D.C., Siegel, E., Wormwood, J.B.: Understanding police performance under stress: insights from the biopsychosocial model of challenge and threat. Front. Psychol. **10**, 1800 (2019)
26. Kermack, W.O., McKendrick, A.G.: Contributions to the mathematical theory of epidemics–I. Bull. Math. Biol. **53**, 33–55 (1991). https://doi.org/10.1007/BF02464423
27. Lai, K., Yanushkevich, S., Shmerko, V.: Intelligent stress monitoring assistant for first responders. IEEE Access **9**, 25314–25329 (2021)
28. Lai, K., Yanushkevich, S., Shmerko, V.: Epidemiology attack at the aircraft carrier Theodore Roosevelt: bridging gaps of the emergency management. J. Defense Model. Simul. Open Excess (2021)
29. Lawless, W.: Risk determination versus risk perception: a new model of reality for human-machine autonomy. Informatics (Basel) **9**(30), 30 (2022)
30. Lazebnik, T., Alexi, A.: Comparison of pandemic intervention policies in several building types using heterogeneous population model. Commun. Nonlinear Sci. Numer. Simul. **107**, 106176 (2022)
31. Lefevr, N., Kanavos, A., Gerogiannis, V.C., Iliadis, L., Pintelas, P.: Employing fuzzy logic to analyze the structure of complex biological and epidemic spreading models. Mathematics **9**, 977 (2021)
32. Lima, L.L., Atman, A.P.F.: Impact of mobility restriction in COVID-19 super spreading events using agent-based model. PLoS ONE **16**(3), e0248708 (2021)
33. Li, J., Tang, J., Wang, D.: Dynamic spreading model of passenger group panic considering official guidance information in subway emergencies. Math. Problems Eng. **2019**, 4691641 (2019)
34. Lin, J., et al.: Trust in the danger zone: individual differences in confidence in robot threat assessments. Front. Psychol. **13**, 601523 (2022)
35. Lyons, J.B., Vo, T., Wynne, K.T., Mahoney, S., et al.: Trusting autonomous security robots: the role of reliability and stated social intent. Hum. Factors **63**(4), 603–618 (2021)

36. Maglione, M.A., et al.: Stress Control for Military, Law Enforcement, and First Responders: A Systematic Review, Santa Monica. RAND Corporation, Calif (2021)
37. Mandrake, L., Doran, G., Goel, A., et al.: Space applications of a trusted AI framework: experiences and lessons learned. In: Proceedings IEEE Aerospace Conference (AERO) (2022). https://doi.org/10.1109/AERO53065.2022.9843322
38. Martcheva, M.: An Introduction to Mathematical Epidemiology. Springer, New York (2015). https://doi.org/10.1007/978-1-4899-7612-3
39. Munoz, S., Iglesias, C.A.: An agent based simulation system for analyzing stress regulation policies at the workplace. J. Comp. Sci. **51**, 101326 (2021)
40. Ojha, R., Ghadge, A., Tiwari, M.K., Bititci, U.S.: Bayesian network modelling for supply chain risk propagation. Int. J. Prod. Res. **56**(17), 5795–5819 (2018)
41. Pare, P.E., Beck, C.L., Başar, T.: Modeling, estimation, and analysis of epidemics over networks: an overview. Annu. Rev. Control. **50**, 345–360 (2020)
42. Pandey, P., Lee, E.K., Pompili, D.: Detection of stress and of its propagation in a team. IEEE J. Biomed. Health Inf. **20**(6), 1502–1512 (2016)
43. Panganiban, A.R., Matthews, G., Long, M.D.: Transparency in autonomous teammates: intention to support as teaming information. J. Cogn. Eng. Decision Making **14**(2), 174–190 (2020)
44. Paoluzzi, M., Gnan, N., Grassi, F., Salvetti, M., Vanacore, N., Crisanti, A.: A single-agent extension of the SIR model describes the impact of mobility restrictions on the COVID-19 epidemic. Sci. Rep. **11**, 24467 (2021)
45. Patil, D., Ansari, M., Tendulkar, D., et al.: A survey on autonomous military service robot. In: International Conference on Emerging Trends in Information Technology and Engineering, India, pp. 1–7 (2020)
46. Peternel, L., Tsagarakis, N., Caldwell, D., Ajoudani, A.: Robot adaptation to human physical fatigue in human-robot co-manipulation. Auton. Robot. **42**, 1011–1021 (2018)
47. Perez, L., Dragicevic, S.: An agent-based approach for modeling dynamics of contagious disease spread. Int. J. Health Geogr. **8**, 50 (2009). https://doi.org/10.1186/1476-072X-8-50
48. Pollak, A., Paliga, M., Pulopulos, M.M., Kozusznik, B., Kozusznik, M.W.: Stress in manual and autonomous modes of collaboration with a cobot. Comp. Hum. Behav. **112**, 106469 (2020)
49. Prichard, J., Bizo, L., Stratford, R.: Evaluating the effects of team-skills training on subjective workload. Learn. Instr. **21**(3), 429–440 (2011)
50. Rebensky, S., Carmody, K., Ficke, C., Carroll, M., Bennett, W.: Teammates instead of tools: the impacts of level of autonomy on mission performance and human-agent teaming dynamics in multi-agent distributed teams. Front. Robot. AI **9**, 782134 (2022)
51. Rodrigues, S., Paiva, J.S., Dias, D., Pimentel, G., Kaiseler, M., Cunha, J.P.S.: Wearable biomonitoring platform for the assessment of stress and its impact on cognitive performance of firefighters: an experimental study. Clin. Pract. Epidemiol. Mental Health CP & EMh **14**, 250 (2018)
52. Scata, M., Stefano, A.D., Corte, A.L., Lio, P.: Quantifying the propagation of distress and mental disorders in social networks. Sci. Rep. **8**, 5005 (2018)
53. Seeber, I., Waizenegger, L., Seidel, S., et al.: Collaborating with technology-based autonomous agents: issues and research opportunities. Internet Res. **30**(1), 1–18 (2020)
54. Stanley, I.H., Hom, M.A., Joiner, T.E.: A systematic review of suicidal thoughts and behaviors among police officers, firefighters, EMTs, and paramedics. Clin. Psychol. Rev. **44**, 25–44 (2016)

55. Tharion, W., et al.: Evolution of physiological status monitoring for ambulatory military applications. In: Matthews, M., Schnyer, D. (eds.) Human Performance Optimization: The Science and Ethics of Enhancing Human Capabilities. Oxford Academic, January 2019. https://doi.org/10.1093/oso/9780190455132.003.0007
56. Ulbrich, S., Menzel, T., Reschka, A., Schuldt, F., Maurer, M.: Defining and substantiating the terms scene, situation, and scenario for automated driving. In: Proceedings IEEE 18th International Conference on Intelligent Transportation Systems, pp. 982–988 (2015)
57. Urizar, O.J., Baig, M.Z., Barakova, E.I., et al.: A hierarchical Bayesian model for crowd emotions. Front. Comput. Neurosci. **10**, 63 (2016)
58. Veld, E.M.J., De Gelder, B.: From personal fear to mass panic: the neurological basis of crowd perception. Hum. Brain Mapp. **36**, 2338–2351 (2015)
59. Wild, J., El-Salahi, S., Esposti, M.D.: The effectiveness of interventions aimed at improving well-being and resilience to stress in first responders: a systematic review. Eur. Psychol. **25**(4), 252–271 (2020)
60. Yan, H., Ang, M., Poo, A.: A survey on perception methods for human-robot interaction in social robots. Int. J. Soc. Robot. **6**(1), 85–119 (2014)
61. Yanushkevich, S., Shmerko, V.: Introduction to Logic Design. CRC Press, Taylor and Francis Group, Boca Raton (2008)
62. Yanushkevich, S.N., Miller, D.M., Shmerko, V.P., Stankovic, R.S.: Decision Diagram Techniques for Micro- and Nanoelectronic Design. CRC Press, Taylor and Francis Group, Boca Raton (2006)
63. Ziegler, C.-N., Golbeck, J.: Models for trust inference in social networks. In: Król, D., Fay, D., Gabryś, B. (eds.) Propagation Phenomena in Real World Networks. ISRL, vol. 85, pp. 53–89. Springer, Cham (2015). https://doi.org/10.1007/978-3-319-15916-4_3
64. Zhang, Q., Busemeyer, J.: A quantum walk model for idea propagation in social network and group decision making. Entropy **23**, 622 (2021)
65. Zobolas, J., Monteiro, P.T. Kuiper, M., Flobak, A.: Boolean function metrics can assist modelers to check and choose logical rules. J. Theor. Biol. **538**, 111025 (2022)
66. Zuck, M.V., Frey, R.J.: The gale encyclopedia of medicine. In: Longe, J.L. (ed.) Gale, Cengage Company, 5th ed. (2015)

AVCA: Autonomous Virtual Cognitive Assessment

Bahar Karimi[1], Soheil Zabihi[2], Amir Keynia[3], Aram Montazami[3], and Arash Mohammadi[1(✉)]

[1] Concordia Institute of Information Systems Engineering (CIISE), Montreal, Canada
b_rimi@live.concordia.ca, arash.mohammadi@concordia.ca
[2] Electrical and Computer Engineering, Concordia University, Montreal, Canada
s_zab@encs.concordia.ca
[3] NOVATEK International, Quebec H4R -2E9, Canada
{amir.keynia,aram.montazami}@ntint.com

Abstract. Fast-paced and ever-growing advances in Artificial Intelligence (AI) and Deep Neural Network (DNN) models have initiated works on autonomous monitoring/screening systems to assess individuals' cognitive state. In conventional cognitive assessment systems, a physician evaluates the mental abilities of the brain by rating the patient's numerical, verbal, and logical responses. Development of an autonomous cognitive assessment system that assist the physician is both of significant importance and a critically challenging task. As a first step towards achieving this objective, in the paper an Automated Virtual Cognitive Assessment (AVCA) framework is proposed that integrates Natural Language Processing (NLP) and hand gesture recognition techniques. The proposed AVCA framework provides individual scores in the seven major cognitive domains, i.e., orientation, attention, language, contractual ability, memory, calculation, and reasoning. More specifically, the AVCA framework is an autonomous cognitive assessment system that receives audio and video signals in a real-time fashion, and performs semantic and synthetic analysis using NLP techniques and DNN models. Real-time video processing engines of the AVCA monitors hand motions and facilitate simpler engagement for visual evaluation. Additionally, we propose an efficient model to facilitate human-machine interactions from speech recognition to text classification. In particular, an unsupervised contrastive learning framework is proposed using Bidirectional Encoder Representations from Transformers (BERT) that outperforms its state-of-the-art unsupervised counterparts achieving an average of 77.84% Sparsman's correlation on standard Semantic Textual Similarity (STS) tasks.

Keywords: Cognitive Assessment · Virtual Assistant · Natural Language Processing · Sentence Embedding · Hand Gesture Recognition

This Project was partially supported by Department of National Defence's Innovation for Defence Excellence & Security (IDEaS), Canada.

1 Introduction

The COVID-19 pandemic has abruptly and undoubtedly changed the world as we knew at the end of the second decade of the 21st century. It is expected for the negative effects of the COVID-19 to continue for years to come. To overcome these long-lasting effects and be prepared for future possible ones, special attention should be devoted to the fact that seniors, aged 60 and over, are more vulnerable during the pandemic era. In addition to higher risk of infections, seniors are also at higher risk of suffering from mental and cognitive issues. Given an expected and alarming population aging in near future, it is crucial and of significant importance to develop innovative and advanced autonomous screening systems for unconstraint environments. Of particular interest to this paper is development of advanced autonomous cognitive screening system via integration of Signal Processing (SP), Artificial Intelligence (AI), and Deep Learning (DL) models. In particular, the focus is on development of an AI-empowered avatar that autonomously performs cognitive screening tests.

Conventionally, cognitive health monitoring is performed manually by physicians, however, there is an urgent and unmet quest to develop autonomous systems for cognitive (mental) health screening. This is of significant importance as in this point in time we are dealing with long lasting effects of the COVID-19 pandemic, among which isolation and mental health issues are devastating. Generally speaking, mental state refers to the current set of emotions and psychological states of a person [1]. It was argued [2] that poor mental health, caused due to excessive stress and/or other negative emotions, can lead to an added emotional and/or economic burden on individuals suffering from mental health problems. Different modalities can be used to perform emotion state monitoring. In brief, emotions can be recognized from modalities such as speech, facial expressions, physiological signals, behavior (gesture/posture), and/or online social media textual posts. Reference [3] reviews commonly used recognition methodologies to perform emotion expressions from facial images. Another modality that has been extensively studied for emotion recognition is through Electroencephalogram (EEG) signals [4]. The concept of EEG-based emotion recognition provides a unique solution with application in Human Machine Interfaces (HMIs) and health care. Reference [5] focused on multimodal emotion recognition integrating EEG signals with other modalities. Reference [6] integrated muscular signals obtained from Electromyography (EMG) with other modalities for the task of human emotion recognition. Finally, in [7], user-generated online social media textual posts are used for development of an emotion recognition system. Early detection of deteriorating mental health, e.g., due to excessive stress and/or anxiety, can prevent conditions such as burnout and depression in highly demanding professions such as nurses and surgeons [8]. In what follows and before presenting contributions of the paper, we briefly outline basics of cognitive state monitoring systems followed by an overview of different cognitive assessment tools.

Cognitive Assessment: Over the years, several solutions have been proposed to monitor mental states, however, mental states have traditionally been monitored using subjective questionnaires. Different questionnaires are developed and applied for various mental states. Such questionnaires, typically, ask about a person's mental health on a regular basis. The response to such questionnaires is typically in rating format (continuous scale) or one of several specified options (discrete scale). Discrete categorization of emotions includes the six fundamental emotions of happiness, sadness, fear, anger, disgust and surprise. In this scale, more complicated emotions (such as exhaustion or anxiety) are viewed as a mixture of these fundamental ones. Other more complex emotions (such as fatigue or anxiety) are considered to be a combination of these basic emotions under this framework [9].

An alternative approach to mental state monitoring via questionnaires is by using audio-visual data generated by individuals. Humans communicate emotions with one another using speech, facial expressions, and body gestures. Hence, using these modalities has been a popular method for emotion state monitoring. Another approach to evaluate mental state is by using neurophysiological signals. For example, in [10] different algorithms are developed exploring neurophysiological correlates of mental state. Reference [11] focused on mental state of pilots and looked at neurophysiological changes corresponding to the mental state transitions. Finally, authors in [12] focused on evaluating mental state of construction workers via neurophysiological signals. These signals reflect the activity of the central and autonomic nervous systems. Central nervous system (CNS) activities can be captured by monitoring electrical, magnetic or hemodynamic activities of the brain. Electroencephalography (EEG) is one such method that captures the electrical activity of the brain using electrodes placed on the scalp surface. EEG has high temporal resolution compared to other neuro-imaging approaches and thus can be used for real-time mental state monitoring [13]. The Autonomic Nervous System (ANS) activity, in turn, regulates bodily functions and can be captured using physiological signals such as Electrocardiogram (ECG), respiration signal, skin temperature, and Galvanic Skin Response (GSR), to name a few. The ANS can be further divided into two categories, i.e., (i) The Parasympathetic Nervous System (PNS), which relaxes the body, and; (ii) The Sympathetic Nervous System (SNS), which is associated with the flight-or-fight response. These two systems impact physiological signals differently and can help distinguish between different mental states [14]. Combining the different signals to create a multi-modal system for mental state monitoring has shown improved performance. This is due to the fact that different signals provide complementary information, as well as added robustness to sensor failure and high noise levels that could be present in individual modalities [25].

A mental state examination provides you with a momentary view of a patient's emotions, thoughts, and behaviour as they are observed. In contrast, a mental status test measures present mental capacity by assessing general appearance, behavior, any unusual or bizarre beliefs and perceptions (e.g., delusions, hallucinations), mood, and all aspects of cognition (e.g., attention, orientation, memory).

There are three major means (types) of assessing a patient's mental status [26]. One type focuses on the patient's capacity to carry out instrumental activities and activities of daily life in order to assess whether or not they have dementia and how severe it is. In the second form of assessment, "screening tests" or "omnibus tests" are used. These quick tests are carried out separately from the patient's history and physical examination. According to [27], the Mini-Mental Status Examination (MMSE) and the Montreal Cognitive Assessment (MoCA) are the most widespread psychometric cognitive screening tests worldwide. It is worth mentioning that MMSE is seemingly more sensitive to severe dysfunctions while the MoCA is more sensitive to mild-to-moderate dysfunctions. Recently, MMSE and MoCA have also been effective in detecting cognition deficits following COVID-19 [28]. The third means of assessing a patient's mental status is by using specific neuropsychological tests that focus on specific domains of cognition. According to [26], these evaluations are performed on patients who exhibit behavioural and cognitive abnormalities, and they should frequently be accompanied by laboratory and neuroimaging tests that can help identify the underlying pathologic process and enable the development of efficient therapeutic and management strategies.

In summary, cognitive assessment is effective to test for cognitive impairment - a deficiency in knowledge, thought process, or judgment. Psychiatrists often perform cognitive testing during the mental status exam, however, when cognitive impairment is suspected, the cognitive assessment can obtain a more detailed analysis by surveying the neuropsychological domains. This detailed investigation of cognition can diagnose major cognitive impairment (i.e., dementia) and mild cognitive impairment, evaluate traumatic brain injuries, help determine decision-making capacity, and survey intellectual dysfunction.

Cognitive Assessment Tools: Cognitive assessment is often performed using clinically validated tools, such as the Mini Mental State Examination (MMSE) [15], the Montreal Cognitive Assessment (MoCA) [16], and Neurobehavioral Cognitive Status Examination (NCSE) or COGNISTAT [17], which are briefly introduced below:

– *Mini Mental State Examination (MMSE):* Perhaps the most extensively used clinical measure for determining the degree of cognitive impairment is the MMSE. The MMSE was created as a bedside measure to assess older patients' cognitive health in clinical settings; it is frequently used in surveys to test for dementia and cognitive impairment. It is quick and simple to manage, has an excellent track record of reliability, although several restrictions have been noted. The MMSE may not detect focal brain damage or mild dementia, and validity may be poor in samples that contain psychiatric patients [29]. The MMSE has a maximum score of 30. Normal is defined as a score of 25 or higher. A score of less than 24 is usually considered abnormal, indicating possible cognitive impairment.

– *Montreal Cognitive Assessment (MoCA):* MoCA was created as a screening tool for patients who have mild cognitive complaints and typically score in the normal range on the MMSE [16]. The final version of the MoCA is a one-page 30-point test administered in 10 minutes. Details on the specific MoCA items are as follows: The short-term memory recall task (5points) involves two learning trials of five nouns and delayed recall after approximately 5 minutes.

Visuospatial abilities are assessed using a clock-drawing task (3 points) and a three-dimensional cube copy (1 point). Multiple aspects of executive functions are assessed using an alternation task adapted from the Trail Making B task (1 point), a phonemic fluency task (1 point), and a two-item verbal abstraction task (2 points). Attention, concentration, and working memory are evaluated using a sustained attention task (target detection using tapping; 1 point), a serial subtraction task (3 points), and digits forward and backward (1 point each). Language is assessed using a three-item confrontation naming task with low-familiarity animals (lion, camel, rhinoceros; 3 points), repetition of two syntactically complex sentences (2 points), and the aforementioned fluency task. Finally, orientation to time and place is evaluated (6 points).

- *Neurobehavioral Cognitive Status Examination (NCSE).* The NCSE or COG-NISTAT [17] is another cognitive screening test of separate subtests assessing major domains of cognitive functions. These are orientation, attention, language (comprehension, repetition and naming), construction, memory, calculation, similarity and judgement. COGNISTAT was found to be sensitive to the cognitive effects of stroke, although there was little discrimination between left-sided and right-sided strokes [20]. COGNISTAT is more sensitive than MMSE to identify cognitive impairment in geriatric inpatients with a variety of medical conditions as well as individuals with brain tumours and cerebrovascular disease. [18]. The sensitivity of MoCA-J and COGNISTAT is thought to be nearly comparable, making them both suitable for detecting Mild Cognitive Impairment (MCI) in Parkinson's disease patients (PD). These outcomes concern patients who have previously received a PD diagnosis [19].

The developers of COGNISTAT claim that their test "represents a new method to quick cognitive testing." This evaluation provides the clinician with a unique profile of the patient's cognitive status by independently assessing numerous domains of cognitive functioning. [17]. The NCSE is a tool that employs an initial screening question for each cognitive domain; if the patient is unable to respond, the tool determines the severity of impairment. If the patient passes the screen, the ability involved is assumed normal, and no further testing is done in that section. Instead of using a single overall score, the points given for correct responses are summed within each category of cognitive ability.

Contributions: In this study, we aim to implement an autonomous system for the evaluation of patients with cognitive impairments. In particular, the main objective is implementing the activities associated with the NCSE assessment system in an autonomous fashion. In achieving this objective, we have formulated the following research questions:

- The first targeted research question is what are the key components of the NCSE assessment system that can be implemented autonomously.
- The second research question is what technologies are needed (available) to be utilized for automated implementation of the identified components.

- The third research question is what are the new methodologies that are needed to be developed (or improved) to provide enhanced performance.

To achieve the aforementioned objective, we made the following contributions:

- Our first contribution is designing a Human Machine Interface (HMI) system, referred to as the Autonomous Virtual Cognitive Assessment (AVCA) framework, which automates cognitive assessment tests. The AVCA framework is developed based on the NCSE concept by identifying its main components across orientation, attention, language, contractual ability, memory, calculation and reasoning dimensions.
- The second contribution of the paper is with regards to the second research question, where advanced NLP methods are identified and integrated with hand gesture recognition techniques to automate cognitive examination.
- The third contribution of the paper follows the aforementioned third research question. In this regard, we propose a new data augmentation technique using unsupervised contrastive learning that improves performance of recent state-of-the-art sentence embedding models. Along the same direction, a Feed Forward Neural network (FNN) model is developed that more accurately evaluates patient's responses in the reasoning section of NCSE.

The remainder of the paper is organized as follows: Sect. 2 analyzes various language models and speech recognition techniques for building a virtual assistant system. In Sect. 3, the AVCA's NLP module, the suggested sentence embedding framework, the COGNISTAT testing tasks utilizing hand gesture detection are described. Experiments and results are presented in Sect. 4, where various tasks of cognitive assessments are explained along with how we implemented their automated version. Finally, Sect. 5 concludes the paper.

2   Materials and Method

Clinically recognized cognitive assessment tools require the presence of trained clinical staff [21] since the presentation of instructions and stimuli must be conducted in a clinically validated, controlled manner for achieving optimal results [22,23]. This constraint makes the frequent and longitudinal assessment difficult to be achieved in/outside the clinical setting [24]. Development of an autonomous cognitive assessment system that replaces the physician is, however, a critically challenging task. As stated previously, the proposed AVCA framework integrates NLP methods with hand gesture recognition to automate cognitive examination as briefly described below.

2.1   Language Modeling

Language Modeling (LM) has been a central task in NLP with the goal of learning a probability distribution over sequences of symbols pertaining to a language. Deep learning models [30], including Long-Short Term Memory networks

(LSTM) [31] and Transformer architectures, are state-of-the-art modeling techniques utilized for different NLP tasks, such as sequence labeling [32], named entity recognition (NER) [33] and part of speech (POS) tagging [34]. Language models developed based on attention mechanism and self-training methods [35] such as ELMo [36], GPT [37], Bidirectional Encoder Representations from Transformers (BERT) [39], XLM [47], and XLNet [38] have brought significant performance gains to the NLP domain. Transformer architecture, which constitutes the main component of such models, uses a self-attention mechanism to capture dependencies among various instances of the input sequence. Self-attention is developed based on a scaled dot-product attention function, mapping a query and a set of key-value pairs to an output. The query (Q), keys (K), and values (V) are learnable representative vectors for the instances in the input sequence with dimensions d_k, d_k, and d_v, respectively. The output of the self-attention module is computed as a weighted average of the values, where the weight assigned to each value is computed by a similarity function of the query and the corresponding key after applying a softmax function. More specifically, the attention values on a set of queries are computed simultaneously, packed together into matrix $\boldsymbol{Q}$. The keys and values are similarly represented by matrices $\boldsymbol{K}$ and $\boldsymbol{V}$. The output of the attention function is computed as follows

$$Attention(\boldsymbol{Q}, \boldsymbol{K}, \boldsymbol{V}) = \text{softmax}\left(\frac{\boldsymbol{Q}\boldsymbol{K}^T}{\sqrt{d_k}}\right)\boldsymbol{V}, \tag{1}$$

where superscript ^{T} denotes transpose of a given matrix. It is also beneficial to linearly project the queries, keys, and values h times with various learnable linear projections to vectors with d_k, d_k and d_v dimensions, respectively, before applying the attention function. On each of the projected versions of queries, keys, and values, the attention function is performed in parallel, resulting in $d_v - dimensional$ output values. These values are then concatenated and once again linearly projected via a FC layer. This process is called Multi-head Self Attention (MSA), which helps the model to jointly attend to information from different representation subspaces at different positions. The output of the MSA module is given by

$$MSA(\boldsymbol{Q}, \boldsymbol{K}, \boldsymbol{V}) = Concat(head_1, \cdots, head_h)\boldsymbol{W}^O,$$
$$\text{where} \quad head_i = Attention(\boldsymbol{Q}\boldsymbol{W}_i^Q, \boldsymbol{K}\boldsymbol{W}_i^K, \boldsymbol{V}\boldsymbol{W}_i^V), \tag{2}$$

where the projections are achieved by parameter matrices $\boldsymbol{W}_i^Q \in \mathbb{R}^{d_{model} \times d_k}$, $\boldsymbol{W}_i^K \in \mathbb{R}^{d_{model} \times d_k}$, $\boldsymbol{W}_i^V \in \mathbb{R}^{d_{model} \times d_v}$, and $\boldsymbol{W}^O \in \mathbb{R}^{hd_v \times d_{model}}$.

Word and Sentence Embedding. To perform autonomous cognitive state analysis, word/sentence embeddings are needed to construct representation of patients' responses (words/sentences) in the form of real-valued vectors to encode the underlying meaning. A contextualized word embedding is a vector representing a word in a special context. Traditional word embeddings such as Word2Vec and GloVe generate one vector for each word, whereas contextualized word embeddings such as ELMo, ULMFiT, and BERT generate a vector for a word depending on the context. Bidirectional LSTM is, typically, used in these contextualised language models. In the ELMo both the forward language model and the backward language model share the same LSTM, therefore, it fails to simultaneously consider the previous and subsequent tokens. Two further contextualised language models are GPT3 and BERT. GPT Transformer also uses constrained self-attention where every token can only attend to context to its left. BERT only uses the encoder blocks from the transformer and employs bidirectional self-attention, whereas GPT3 only employs the decoder blocks from the transformer and employs limited self-attention, where each token can only pay attention to the context to its left. GPT3 is a potent text generation tool, while BERT can be quickly fine-tuned on a particular downstream task with a limited number of labels. In the proposed AVCA framework, we utilized the BERT architecture, which has the following variants:

- *BERT Architecture:* BERT [39] is a multi-layer bidirectional Transformer encoder that relies entirely on attention mechanisms [40], with no recurrence or convolutions. Unlike most other language models, BERT's transformer encoder reads the entire sequence of words at once, randomly masks words in the sentence and predicts them. Bert includes two training processes; pre-training and fine-tuning. The pre-trained parameters are first used to initialize the BERT model for fine-tuning, and labelled data from the downstream tasks is used to fine-tune each parameter. Despite being initialized with the same pre-trained parameters, each downstream task has its own fine-tuned models.
- *RoBERTa Architecture:* RoBERTa [35] is an enhanced method for training BERT models that can match or even outperform the effectiveness of existing post-BERT techniques. With more data, longer sequences, and larger batches, the model is trained over a longer period of time. Additionally, the objective for predicting the following sentence is dropped, and the masking pattern that was used to mask the training data is dynamically altered.
- *Sentence-BERT Architecture:* To create a fixed-sized sentence embedding, SBERT [41] adds a pooling operation to the output of RoBERTa and BERT. The mean of all output vectors is calculated as the default pooling approach. Siamese and triplet networks [42] are developed to update the weights in BERT/RoBERTa such that the resultant sentence embeddings are semantically meaningful and can be compared with cosine-similarity.

2.2   Automatic Speech Recognition

Automatic Speech Recognition (ASR) allows for the transcription of spoken language into text and is intended to facilitate natural human-machine interaction. All experiments are performed based on the two following open-source ASR models including wav2vec2 [56] and HuBERT [57].

- *The Wav2Vec2* is a transformer-based self-supervised model that encodes speech audio using a multi-layer Convolutional Neural Network (CNN) as

Table 1. Accuracy of wav2vec2 and HuBERT on 100 audio files.

Wav2vec2 (US accent)	Wav2vec2 (UK accent)	HuBERT (US accent)	HuBERT (UK accent)
94.13%	93.44%	90.60%	91.12%

a feature extractor before masking the spans of the resulting latent speech representations. After that, a Transformer network receives the latent representations and creates contextualised representations. The model is then fine-tuned on labelled data using the Connectionist Temporal Classification (CTC) [44] algorithm loss to be used for downstream speech recognition tasks. The wav2Vec2 [56] base model is Wav2Vec2-Base-960h, which has been fine-tuned over 960 h of Librispeech [43] on 16 kHz sampled speech audio.

- *The HuBERT* uses an offline k-means clustering step and learns the structure of spoken input by predicting the right cluster for masked audio segments. The HuBERT [57] progressively improves its learned discrete representations by alternating between clustering and prediction. The predictive loss is applied over only the masked regions, forcing the model to learn good high-level representations of unmasked inputs in order to infer the targets of masked ones correctly.

It is worth mentioning that Hsu *et al.* [57] showed that the HuBERT outperforms wav2vec2 in most scenarios. A test have been conducted on these two ASR models with 100 audio files as input with US and UK accents. Each audio file stated one of the 100 most well-known names in the world. Each text generated by these ASR models were compared to their corresponding source text using string matching. Similarity percentages of at least 60% were deemed to be acceptable answers. This data was chosen because names are prone to more errors in ASR models. The results are illustrated in Table 1. Audio recordings of cities were also evaluated. Wav2vec2's average accuracy was around 70% while Hubert's average accuracy was close to 80%.

This approach was also tested on audio files of cities. Hubert's average accuracy was close to 80%, while wav2vec2's average accuracy was about 70%.

3   The AVCA Framework

The proposed AVCA framework is inherently a Virtual Assistant (VA) [48], that provides individual scores for major domains of Cognitive Assessments. The proposed AVCA frameworks consists of four major modules, i.e., the Human Machine Interface (HMI), NLP, and Sentence Embedding (SE) modules, together with the gesture recognition module, which are presented below.

3.1   Dataset

To train the proposed unsupervised model, we used a dataset consisting of 1 million sentences from English Wikipedia and the combination of MNLI and SNLI datasets [53]. The MNLI and SNLI datasets, which are described below, contain $570K$ and $433K$, respectively, human-written English sentence pairs manually labeled for balanced classification with the labels entailment (one sentence that is absolutely true), contradiction (one sentence that is definitely false), and neutral (one sentence that might be true). The dataset can be denoted as (x_i, x_i^+, x_i^-) where x_i is the premise, x_i^+ and x_i^- are entailment and contradiction sentences. *The Stanford Natural Language Inference (SNLI) Dataset* [54], is a collection of labeled sentence pairs, which is created based on image captioning. The SNLI consists of $570K$ pairs of sentences allowing development of deep learning-based lexicalized classifiers for natural language inference tasks. Amazon Mechanical Turk is used for collecting the SNLI dataset, where in each task a premise scene is presented to the human annotator. The annotator's task is to hypothesizes there sentences, i.e., one entailment, one neutral, and one contradiction. The premises are obtained from a pre-existing corpus, i.e., the Flickr30k corpus consisting collection of approximately 160k captions (corresponding to about 30k images). The released dataset, explicitly the train, development, and test sets. The development and the test sets each consists of $10k$ examples, with each original ImageFlickr caption only appearing in one of the aforementioned three sets.

The Multi-Genre Natural Language Inference (MNLI) Dataset [55], is a widely utilized dataset for evaluation and design of deep learning models for the task of sentence understanding. The MNLI consists of $433k$ examples, making it one the largest datasets that is accessible for sentence understanding. More specifically, the MNLI contains data from ten different spoken and written English genres, which allows for evaluation of new models incorporating the full complexity of English language. The approach used for collection of the MNLI dataset is similar in nature to the one used for construction of the SNLI. Each sentence pair is created by first identifying a sentence as the premise from a pre-existing text, which is called the source. Afterwards, a new sentence is composed to be paired with the premise sentence, as a hypothesis, with the help of a human annotator. Based on development examples, it seems that while the collected hypotheses are not always complete sentences, they are fluent and correctly spelled. The MNLI dataset is available in two formats, i.e., JSON Lines (jsonl) similar to SNLI and tab separated text. Several fields are specified for each example, including unique identifiers for the pair and prompt, and premise and hypothesis strings. The distributed MNLI dataset explicitly provides, train, development, and test sets, with no premise sentence occurring in more than one set. The development and test sets each consists of 20, 000 examples, which include 2, 000 examples randomly selected from each of the ten genres.

For development of the reasoning module of the proposed framework, an in-house dataset is utilized, which is described below:

Reasoning Dataset: An in-house dataset containing 830 responses from various patients was used to construct the evaluation framework. The responses were collected by a physician in the in-person cognitive assessment test and the same judgment task question was asked to each participant. Using our proposed embedding module we transferred the test data into numerical vectors. For the purpose of training we used 95% of the responses as training data and used the remaining 5% as test data.

3.2   HMI, NLP and SE Modules

Patients' audio responses are initially recorded and transformed into text using the ASR Module. Two significant NLP techniques are then employed to process the received text

- *String Matching:* When patients are required to repeat a sentence as part of the cognitive test, this method is employed. The fewest number of symbol substitutions, insertions, and deletions required to change one word into the other is known as Levenshtein Distance (LD). In our implementation, the difference between the patient's answer and the correct response is measured with LD. To adjust for inaccuracies in the speech-to-text module, a difference of 30% or less is regarded as a correct response.
- *Semantic Similarity:* To deal with other test domains where a more complex module is needed to compare phrases semantically as opposed to lexicographical similarity, semantic similarity is used. The semantic similarity of two input text phrases is determined by comparing the cosine similarity of their embeddings. As one of the most popular similarity metrics, the cosine similarity calculates the angle between two non-zero vectors and measures how similar they are using this angle.

The semantic similarity unit of the AVCA framework receives the audio signal recorded as the patient's response. The pre-processing unit will convert the continuous audio signal to text, which is the fed to a BERT model followed by a pooling layer to form the embedding vector. The NLP model compares the embedding of the patient's response with the embedding of its corresponding correct response stored in the database. A cosine similarity unit will then evaluate the semantic similarity of the two embeddings.

The sentence embedding module of the proposed AVCA framework is designed based an unsupervised contrastive learning framework. The contrastive learning model takes two sentences as positive samples in the training phase and treat the other sentences as negative samples. The main idea is to train the representation layer of the model by a contrastive loss objective function [49] to pull closer representations of the same class field (i.e., original sentence and its augmented one(s) called positive samples) and separate them from the rest of the sentences. In this regard, data augmentation is needed, which refers to a series of techniques that create simulated data out of a set of existing data. The predictions of the model ought to be invariant to the modest changes that are frequently present in this simulated data. By randomly rearranging the specific linguistic forms, the regularisation technique known as data augmentation can help prevent over-fitting by minimising the learning rate of erroneous correlations. Numerous regularisation strategies have been explored, including dropout [50] or weight penalties [51]. Augmentations are grouped into symbolic or neural methods [52]. While neural augmentations employ a DNN trained on a separate task to augment data such as Generative Data Augmentation (GDA), symbolic approaches use rules or discrete data structures to produce synthetic instances, such as substituting words or phrases to create augmented examples. Utilizing dropout noise is another method that is regarded as a form of data augmentation. In this method, the positive pairs are one sentence repeated twice, the only difference between their embeddings is in their dropout masks [45]). None of the discrete augmentations outperform dropout noise, and further data augmentation methods like crop, word deletion, and replacement would have a negative impact on performance. In accordance with the concept we provided before, GDA is a sort of neural data augmentation that entails creating indistinguishable text paragraphs for tasks like summarization. Pegasus [46] is a sequence-to-sequence model designed for abstractive text summarization using gap-sentence generation as a pre-training target. Important sentences are removed/masked from an input document in the PEGASUS, and the remaining phrases are then created as one output sequence, much like an extractive summary.

We propose to use Pegasus paraphrasing as data augmentation technique, which maintains the semantic meaning of each input sample in its process. Pre-trained checkpoints of BERT [39] or RoBERTa [35]) is used to generate the embedding of original and augmented sentences. Consequently, the original and augmented sentences are fed as inputs to the encoder with two different dropout masks (as in SimCSE [45].) Dropout masks and attention probabilities (default $p = 0.1$) are applied to fully-connected layers in the Transformer's standard training [40].

By taking the above representation as the sentence embedding an unsupervised model is trained on 1 Million sentences from English Wikipedia and the combination of MNLI and SNLI datasets [53], which are a collection of $570K$ and $433K$ human-written English sentence pairs manually labeled for balanced classification with the labels entailment (one sentence that is absolutely true), contradiction (one sentence that is definitely false), and neutral (one sentence that might be true). The dataset can be denoted as (x_i, x_i^+, x_i^-) where x_i is the premise, x_i^+ and x_i^- are entailment and contradiction sentences. However, as our approach is in an unsupervised fashion, we only use the first sentences to feed to our framework. Feeding the NLI datasets to our unsupervised contrastive learning framework without performing paraphrase augmentation on the second sentence of the positive pairs results in no appreciable improvement.

By grouping semantically similar neighbours together and separating unrelated ones, contrastive learning seeks to acquire effective representation. We adopt a cross-entropy target with in-batch negatives [60] while adhering to the contrastive framework [61]. Given an unlabelled sentence x_i, its embedding is

Table 2. Sentence embedding performance (Spearman's correlation) on STS tasks.

Model	STS12	STS13	STS14	STS15	STS16	STS-B	SICK-R	Avg
			Unsupervised models					
GloVe embeddings(avg.)	55.14	70.66	59.73	68.25	63.66	58.02	53.76	61.32
BERT$_{base}$ (first-last avg.)	39.70	59.38	49.67	66.03	66.19	53.87	62.06	56.70
BERT$_{base}$-flow	58.40	67.10	60.85	75.16	71.22	68.66	64.47	66.55
BERT$_{base}$-whitening	57.83	66.90	60.90	75.08	71.31	68.24	63.73	66.28
IS-BERT$_{base}$	56.77	69.24	61.21	75.23	70.16	69.21	64.25	66.58
CT-BERT$_{base}$	61.63	76.80	68.47	77.50	76.48	74.31	69.19	72.05
SimCSE-BERT$_{base}$ [45]	68.40	82.41	74.38	80.91	78.56	76.85	72.23	76.25
AVCA-BERT$_{base}$	**71.43**	**82.19**	**76.10**	**83.78**	**78.65**	**79.27**	**73.45**	**77.84**

formed as $h_i^z = f(x_i, z)$ with random dropout mask z. The training objective, therefore, becomes

$$l_i = -\log \frac{e^{sim(h_i^{z_i}, f(p(x_i), \hat{z}))/\tau}}{\sum_{j=1}^{N} e^{sim(h_i^{z_i}, h_j^{z_j})}}, \tag{3}$$

where $p(x_i)$ is the augmented version of x_i using Pegasus, τ is a temperature hyperparameter and $sim(h_i, h_j)$ is the cosine similarity $\frac{h_1^\top h_2}{||h_1|| \cdot ||h_2||}$.

The proposed model is illustrated in Fig. 1. The proposed AVCA's sentence embedding model is thoroughly assessed using seven standard Semantic Textual Similarity (STS) tasks, i.e., STS 2012-2016 [62–66]; STS Benchmark [67], and SICK-Relatedness [68]. SentEval is a toolkit for evaluating sentence representations, which measures the relatedness of two sentences based on the cosine similarity of their two representations. The SICK relatedness (SICK-R) task trains a linear model to output a score from 1 to 5 indicating the relatedness of two sentences. The evaluation metric can be Pearson's or Spearman's correlation. It is claimed [69] that Spearman correlation, which measures ranks rather than actual scores, is more suitable for evaluating sentence embeddings. Table 2 provides an overview of the STS task comparison results, where the AVCA model is compared to several of its counterparts. AVCA improves the results on almost all STS tasks. This improvement can be significantly important in semantic similarity related tasks. The use of data augmentation techniques during the production of training data has allowed AVCA to better evaluate the similarity of sentences with similar semantic meaning while having varied vocabularies by extracting more significant elements from the underlying text. Our findings are based on unsupervised models, but we anticipate seeing improved outcomes when supervised models are applied with data augmentation techniques.

3.3   Gesture Recognition Module

Two domains of NSCE, i.e., language comprehension and constructional ability, involve actual interactions with physical objects. In one case, the participants is asked to create an identified shape using different red and while rectangular and triangular blocks. In the other case, different objects are placed in front of the participant after which basic pre-defined tasks such as picking up a pen are prompted. To autonomously conduct these components of the NSCE, in the proposed AVCA framework, hand gesture recognition is utilized where patients can pick up virtual objects and recreate patterns through a live video stream.

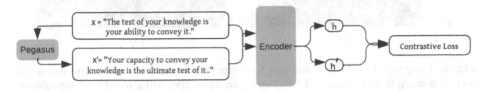

Fig. 1. Sentence encoder is given each sentence along with its Pegasus-augmented equivalent. The input vectors h and h' are regarded as positive pairs for the model with different dropout masks. While other in-batch vectors are viewed as negative instance.

Language Comprehension. In the language comprehension section, patients are asked to use objects in specific ways. An HMI system is developed that allows users to perform the following requested virtual actions: (The <Object> could be keys, coins, pens, or any other defined object in the HMI system).

- Hand me the <Object>: Action detection is based on collision detection between the specified object and the static hand on the screen. To achieve success for this implementation, there were many considerations such as validating the collision point between the objects and the static hand.
- Pick up the <Object>: By confirming that the object is correctly grasped and examining its height on the screen, the pick-up motion is recognized.
- Point to the <Object>: While pointing to the floor detection is based on the angle of the index finger on the screen for both the left and right hands, pointing to nose detection uses face detection, as shown in Fig. 2. The distance between the index finger landmark and the nose landmark is used to identify this action. If this interval is less than the defined threshold for more than the specified time threshold (800 ms), the action is considered complete.

Constructional Ability. For constructional ability section, the patients need to replicate the patterns displayed on the screen using virtual red and white tiles, as shown in Fig. 2. All the game control commands are identified as different hand movements. Hand tracking is used to detect hand movements.

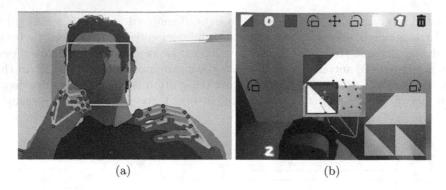

(a)   (b)

Fig. 2. Language Comprehension and Constructional Ability (a) Pointing to nose is detected using face detection, (b) Patients can move tiles using hand-tracking system.

4   Experiments and Results

We put the proposed AVCA cognitive assessment framework into practice in accordance with NSCE. In particular, the three assessments mentioned above, all contain some form of orientation and attention task. These tasks involve asking the patient to repeat a set of phrases or numbers, or in the case of orientation, to state their name or the time. String matching is the sole module used in these tasks. Other tasks, however, require more advanced algorithms. The descriptions of these tasks and our proposed algorithms are presented below.

4.1   Language - Speech Sample

The patients are given 30 seconds to describe what they observe in a given picture throughout this activity. After preprocessing the response, if the number of words or verbs in the response is not enough, a recording will be played asking the patients to say more. Otherwise, the text's words are compared with a set of keywords. If the similarity percentage of each word and its corresponding keyword is more than or equal to 70%, we assume that the patient said the right word. After all the iterations, the similarity of all the responses concatenated and a correct key answer is computed using the AVCA's NLP module. An output percentage of more than or equal to 70 percent is considered a correct answer. The timing of the whole process is also computed and shown.

The speech sample function has been tested with several candidates. The results were promising. All the recorded voices were saying the correct description of the picture, and they all resulted in a semantic similarity percentage of sixty or more with the key sentence. Therefore, 70% is thought to be a suitable threshold for the correct response. Based on the test results, the pronouns of the sentences (e.g., person, boy, girl, etc.) should not be included in the keywords because it would decrease the similarity percentage. Furthermore, three key sentences are added as parameters that are all the same but have three different pronouns (person, boy, girl). In the end, the patient's response will be compared with all the key sentences and the maximum similarity percentage will be selected. One issue with the ASR is that occasionally the converted text sounds correct but the spelling is incorrect. Therefore, the recorded words should also be phonetically compared with the keywords, and if they are phonetically the same, the corresponding word's spelling should be corrected.

4.2 Memory

In this task, the patient is asked to say four words that were mentioned to earlier in the assessment. First, the patient is given a time slot to say the four words. The patient's voice is recorded and converted to text using the AVCA's ASR module. Then, each recorded word is compared with its corresponding keyword using string matching. If the word did not pass the string matching step, it will be compared phonetically. Correct and incorrect responses are stored, and the patient will earn three points for every correct word. For every keyword that was missed by the patient, its category is given to the patient. If the patient remembered that specific keyword or any other unsaid keywords, the patient is given two scores for each of the correct ones. Otherwise, the patient is presented with three words, and one of them is correct. One point is awarded if the patient selects the right option. Otherwise, no score. In the end, the program returns a dictionary, which contains all the keywords, the patient's score for every keyword and the incorrect responses.

4.3 Calculation

In this task, the patient is asked a series of arithmetic calculations. The patient is given a 20-second time slot. In the traditional test, the clinician should note the precise moment the patient provided the right response. This time should, however, be automatically stored in the automatic version. Additionally, it should be noted that the patient might respond incorrectly a few times before responding correctly. The timestamps for all incorrect numbers should be kept. As mentioned in Sect. 2, wav2vec2 is fine-tuned on labeled data with the Connectionist Temporal Classification (CTC), which is a character-based algorithm. During the training phase, it can demarcate each character of the transcription in the speech automatically, therefore, the time-frame alignment is not required between the audio signal and transcription. As a result, in order to save a lot of labelling effort, it does not need to know the time-stamp of each word.

To add time-stamps to the ASR's output tokens, we proposed two solutions. One is to create a real-time speech recognition system using wav2vec2. However, the issue is that while wav2vec2 is trained with a 16khz sampling rate, we record patients' voices at 96khz for higher accuracy. We then turned to our second solution in order to get a higher level of precision. To be able to store the time in the process of speech recognition, we intercepted the wav2vec2 model and added a layer before the final transcription to calculate the time-stamp of each word. From there, we compute the time period of each sample, iterate through the samples while the ASR model is processing and calculate the time each letter was said. To compute the time period of each sample, the total number of samples generated by the ASR's processor is divided by the sampling rate (the number of audio samples recorded per second), which in our case is 16,000. After that, if a letter was detected by the ASR's model iterating through the samples, the letter and its associated time-stamp are saved. Additionally, wav2vec2 output spells out numerals rather than writing them in a numerical style. We created an algorithm to extract all the numerals from a given text in order to preserve all of the patient's wrong responses prior to saying the proper response.

4.4   Reasoning

This task is divided into two sections, i.e., similarity and judgment. In the case of similarities, the patient is asked to state how two items are alike, such as a train and a bicycle. This checks the patients' abstract reasoning. In contrast, judgment assesses the Patients' understanding of the consequences of certain actions, where a set of questions like "What would you do if you found a stamped envelope on the sidewalk?" are asked. As there are a set of correct responses for these questions, an evaluation framework is developed that predicts the score associated with patients' responses as follows:

- *ML-based Evaluation:* Various ML models were used to construct the evaluation framework. The framework based on Support Vector Machine (SVM) produced the highest accuracy of 85.7% compared to logistic regression with 80.9% accuracy. The proposed framework tokenizes the responses after being pre-processed. The pre-processed tokenized responses are then vectorized into n-gram integer vectors using TF-IDF features, to use as the input to the SVM-based classifier model. TFIDF evolved from IDF, which proposes that the frequency of a term in a given document should have an inverse relationship with its associated weight. Equation 4 is the classical formula of used for term weighting [58]. The following TF-IDF formulation is then utilized

$$w_{ij} = t f_{ij} \times \log \left(\frac{N}{df_i} \right), \tag{4}$$

where $w_{i,j}$ is the weight for term i in document j, N is the number of documents in the collection, $t f_{i,j}$ is the term frequency of term i in document j and df_i is the document frequency of term i in the collection.
- *FNN-based Evaluation:* In this scenario, we used a Feed-forward Neural Network (FNN) as the classification module of the AVCA to assess patients' responses. The model contains an input layer, which takes the embedding vector of patients' responses, followed by two hidden layers with $ReLU(\cdot)$ and $tanh(\cdot)$ activation functions coupled with dropout layers to mitigate the risk of over-fitting. Sigmoid is used as the last layer that generates a floating-point number for each sample. We used the grid search method to fine tune the hyper parameters of the FNN. By using the checkpoint callback method, we ensured that the final model is having the minimum validation loss. Hyper

Table 3. Performance comparison based on different classification models.

Model	Accuracy	F1
ktrain [59] + RoBERTa	88%	83%
hline Support Vector Machine + TF-IDF	90%	88%
AVCA FNN + RoBERTa	**93%**	**92%**

parameters were tuned using batch sizes of 12, 32, 64 and 128 with 100 epochs. To evaluate the performance of the final model, a hard cutoff threshold of 0.5 is applied on the model outputs for test samples. Since our data is imbalanced, focal loss is used, which assigns more weights to hard or easily misclassified examples or positive and negative examples. The model is trained by minimizing the focal loss which is given by

$$L(y, \hat{p}) = -\alpha y(1 - \hat{p})\gamma \log(\hat{p}) - (1 - y)\hat{p}\gamma \log(1 - \hat{p}) \tag{5}$$

where $y \in 0, 1$ is a binary class label, $\hat{p} \in [0, 1]$ is an estimate of the probability of the positive class, γ is the focusing parameter that specifies how much higher-confidence correct predictions contribute to the overall loss, and α is a hyper-parameter that governs the trade-off between precision and recall. Table 3 illustrates results based on different models.

5   Conclusion

In this paper, we proposed an AI-powered autonomous system, referred to as the AVCA framework, for cognitive assessment based on the NCSE assessment model. The proposed AVCA framework autonomously provides individual scores in the seven major cognitive domains, i.e., orientation, attention, language, contractual ability, memory, calculation, and reasoning, by integrating NLP and gesture recognition techniques. In this context, the AVCA framework incorporates innovative ideas to autonomously perform the "Calculation" and "Reasoning" tasks. For the former category (i.e., the calculation task), we intercepted an advanced ASR system to extract the time-stamp of every spoken word as well as the patients' numeral responses. For the latter (i.e., the Reasoning task), a classification module is implemented to evaluate patients' responses. Finally, to improve AVCA's evaluation on patients' responses, a novel unsupervised contrastive learning mechanism is proposed based on the BERT model that outperforms state-of-the-art representation solutions. As a direction for future research, one should target incorporation of more advanced data augmentation techniques to increase the precision of the AVCA-BERT$_{base}$. In order to enhance the performance of other sentence embedding models, such as RoBERTa$_{base}$, our data augmentation technique may be used in the context of unsupervised or supervised contrastive learning. Another fruitful direction for future research is the use of a multi-class classification model to more precisely assess patients' responses in the reasoning component.

Conflicts of Interests/Competing Interests. The authors declare no competing interests.

References

1. Jung, H.T., et al.: Remote assessment of cognitive impairment level based on serious mobile game performance: an initial proof of concept. IEEE J. Biom. Health Inf. **23**(3), 1269–1277 (2019)
2. Greenberg, P.E., Fournier, A.-A., Sisitsky, T., Pike, C.T., Kessler, R.C.: The economic burden of adults with major depressive disorder in the united states (2005 and 2010). J. Clin. Psychiatry **76**(2), 155–162 (2015)
3. Canal, F.Z., et al.: A survey on facial emotion recognition techniques: a state-of-the-art literature review. Inf. Sci. **582**, 593–617 (2022)
4. Li, X., et al.: EEG based emotion recognition: a tutorial and review. ACM Comput. Surv. **55**(4), 1–57 (2023)
5. Zhang, J., Yin, Z., Chen, P., Nichele, S.: Emotion recognition using multi-modal data and machine learning techniques: a tutorial and review. Inf. Fusion **59**, 103–126 (2020)
6. Hassan, M.M., et al.: Human emotion recognition using deep belief network architecture. Inf. Fusion **51**, 10–18 (2019)
7. Yousaf, A., et al.: Emotion recognition by textual tweets classification using voting classifier (LR-SGD). IEEE Access **9**, 6286–6295 (2021)
8. Dimou, F.M., Eckelbarger, D., Riall, T.S.: Surgeon burnout: a systematic review. J. Am. Coll. Surg. **222**(6), 1230 (2016)
9. Ekman, P., et al.: Universals and cultural differences in the judgments of facial expressions of emotion. J. Pers. Soc. Psychol. **53**(4), 712 (1987)
10. Ahn, S., et al.: Exploring neuro-physiological correlates of drivers, mental fatigue caused by sleep deprivation using simultaneous EEG, ECG, and fNIRS data. Front. Neurosci. **10**, 1–14 (2016)
11. Han, S.Y., Kwak, N.S., Oh, T., Lee, S.W.: Classification of Pilots. Mental States using a Multimodal Deep Learning Network. Biocybernetics Biomed. Eng. **40**(1), 324–336 (2020)
12. Cheng, B., et al.: Measuring and computing cognitive statuses of construction workers based on electroencephalogram: a critical review. IEEE Trans. Comput. Soc. Syst. **9**(6), 1644–1659 (2022)
13. Borghini, G., Astolfi, L., Vecchiato, G., Mattia, D., Babiloni, F.: Measuring neurophysiological signals in aircraft pilots and car drivers for the assessment of mental workload, fatigue and drowsiness. Neurosci. Biobehav. Rev. **44**, 58–75 (2014)
14. Castaldo, R., Melillo, P., Bracale, U., Caserta, M., Triassi, M., Pecchia, L.: Acute mental stress assessment via short term HRV analysis in healthy adults: a systematic review with meta-analysis. Biomed. Signal Process. Control **18**, 370–377 (2015)
15. Folstein, M.F., Folstein, S.E.: Mini-Mental State" a practical method for grading the cognitive state of patients for the clinician. J. Psychiatric Res. **12**(3), 189–198 (1975)
16. Nasreddine, Z.S., et al.: The Montreal cognitive assessment, MoCA: a brief screening tool for mild cognitive impairment. J. Amer. Geriatrics Soc. **53**, 695–699 (2005)

17. Kiernan, R.J., Mueller, J., Langston, J.W., Van Dyke, C.: The neurobehavioral cognitive status examination: a brief but differentiated approach to cognitive assessment. Ann. Intern. Med. **1987**(107), 481–5 (1987)
18. Fields, S.D., Fulop, G., Sachs, C.J., Strain, J., Fillit, H.: Usefulness of the neurobehavioral cognitive status examination in the hospitalized elderly. Int. Psychogeriatr. **4**, 93–102 (1992)
19. Murakami, H., Fujita, K., Futamura, A., et al.: The Montreal Cognitive Assessment (MoCA) and Neurobehavioral Cognitive Status Examination (COGNISTAT) are useful for screening mild cognitive impairment in Japanese patients with Parkinson's disease. Neurol. Clin. Neurosci. **1**, 103–8 (2013)
20. Osmon, D.C., Smet, I.C., Winegarden, B., Gandhavadi, B.: Neurobehavioral cognitive status examination: its use with unilateral stroke patients in a rehabilitation setting. Arch. Phys. Med. Rehabil. **73**, 414–418 (1992)
21. Tangalos, E.G., et al.: The mini-mental state examination in general medical practice: clinical utility and acceptance. Mayo Clinic Proc. **71**(9), 829–837 (1996)
22. Brouillette, R.M., et al.: Feasibility, reliability, and validity of a smartphone-based application for the assessment of cognitive function in the elderly. PLoS One **8**(6) (2013). Art. no e65925
23. Timmers, C., Maeghs, A., Vestjens, M., Bonnemayer, C., Hamers, H., Blokland, A.: Ambulant cognitive assessment using a smartphone. Appl. Neuropsychol. **21**(2) (2013). Art. no. e112197
24. Schweitzer, P., et al.: Feasibility and validity of mobile cognitive testing in the investigation of age-related cognitive decline. Int. J. Methods Psychiatric Res. **26**(3) (2017). Art. no. e1521
25. Koelstra, S., et al.: Deap: a database for emotion analysis; using physiological signals. IEEE Trans. Affect. Comput. **3**(1), 18–31 (2011)
26. Finney, G.R., Minagar, A., Heilman, K.M.: Assessment of mental Status. Neurol. Clin. **34**(1), 1–16 (2016)
27. Tsoi, K.K., et al.: Cognitive tests to detect dementia: a systematic review and meta-analysis. JAMA Internal Med. **175**, 1450–1458
28. Daroische, R., et al.: Cognitive impairment after COVID-19 - a review on objective test data. Front. Neurol. **12**, 1238 (2021)
29. Tombaugh, T.N., McIntyre, N.J.: The mini-mental state examination: a comprehensive review. J. Amer. Geriatrics Soc. **40**(9), 922–935 (1992)
30. Goyal, A.K., Metallinou, A., Matsoukas, S.: Fast and scalable expansion of natural language understanding functionality for intelligent agents. In: Proceedings of the 2018 Conference of the North American Chapter of the Association for Computational Linguistics: Human Language Technologies, Volume 3 (Industry Papers). Association for Computational Linguistics (2018)
31. Gers, F.A., Schmidhuber, J., Cummins, F.: Learning to forget: continual prediction with lstm (2000)
32. Chung, C.G., Cho, K., Bengio, Y.: Empirical evaluation of gated recurrent neural networks on sequence modelling. In NIPS 2014 Workshop on Deep Learning (2014)
33. Chiu, J.P.C., Nichols, E.: Named entity recognition with bidirectional lstm-cnns. arXiv preprint arXiv:1511.08308 (2015)
34. Huang, Z., Xu, W., Yu, K.: Bidirectional lstm-crf models for sequence tagging. arXiv preprint arXiv:1508.0199 (2015)
35. Liu, Y., et al.: Roberta: a robustly optimized bert pretraining approach. arXiv preprint arXiv:1907.11692 (2019)
36. Peters, M., et al.: Deep contextualized word representations. In: North American Association for Computational Linguistics (NAACL) (2018)

37. Radford, A., Narasimhan, K., Salimans, T., Sutskever, I.: Improving language understanding with unsupervised learning. Technical report, OpenAI

38. Yang, Z., Dai, Z., Yang, Y., Carbonell, J., Salakhutdinov, R., Le, Q.V.: Xlnet: generalized autoregressive pretraining for language understanding. arXiv preprint arXiv:1906.08237(2019)

39. Devlin, J., Chang, M.-W., Lee, K., Toutanova, K.: BERT: pre-training of deep bidirectional Transformers for language understanding. In: Proc. 2019 Conf. North Am. Chapter Assoc. Comput. Linguist. Hum. Lang. Technol., vol. 1, NAACL-HLT 2019, Minneapolis, MN, USA (2019), pp. 4171–4186. Google-AI Language, June 2–7, 2019

40. Vaswani, A., et al.: Attention is all you need. In: Advances in Neural Information Processing Systems, pp. 6000–6010 (2017)

41. Reimers, N., Gurevych, I.: Sentence-BERT: Sentence embeddings using siamese BERT-Networks. DOI (2019)

42. Schroff, F., Kalenichenko, D., Philbin, J.: FaceNet: a unified embedding for face recognition and clustering (2015)

43. Panayotov, V., Chen, G., Povey, D., Khudanpur, S.: Librispeech: an ASR corpus based on public domain audio books. In: Proceedings of ICASSP, pp. 5206–5210. IEEE (2015)

44. Graves, A., Fernández, S., Gomez, F.: Connectionist temporal classification: labelling unsegmented sequence data with recurrent neural networks. In: Proceedings of ICML (2006)

45. Gao, T., Yao, X., Chen, D.: SimCSE: simple contrastive learning of sentence embeddings. In: International Conference on Empirical Methods in Natural Language Processing, pp. 6894–6910 (2021)

46. Zhang, J., Zhao, Y., Saleh, M., Liu, P.J.: Pegasus: pre-training with extracted gap-sentences for abstractive summarization. ICML (2020)

47. Lample, G., Conneau, A.: Crosslingual language model pretraining. arXiv preprint arXiv:1901.07291 (2019)

48. Gondala, S., Verwimp, L., Pusateri, E., Tsagkias, M., Van Gysel, C.: Error-Driven Pruning of Language Models for Virtual Assistants. arXiv:2102.07219 (2021)

49. Chopra, S., Hadsell, R., LeCun, Y.: Learning a similarity metric discriminatively, with application to face verification. In: CVPR (2005)

50. Srivastava, N., Hinton, G.E., Krizhevsky, A., Sutskever, I., Salakhutdinov, R.: Dropout: a simple way to prevent neural networks from overfitting. J. Mach. Learn. Res. (JMLR) 15(1), 1929–1958 (2014)

51. Kukacka, J., Golkov, V., Cremers, D.: Regularization for deep learning: a taxonomy. arXiv preprint arXiv:1710.10686 (2017)

52. Shorten, C., Khoshgoftaar, T.M., Furht, B.: Text data augmentation for deep learning. J. Big Data 8(1), 1–34 (2021)

53. Nie, Y., Williams, A., Dinan, E., Bansal, M., Weston, J., Kiela, D.: Adversarial NLI: a new benchmark for natural language understanding. Association for Computational Linguistics (ACL), pp. 4885–4901 (2020)

54. Bowman, S.R., Angeli, G., Potts, C., Manning, C.D.: A large annotated corpus for learning natural language inference. In: Conference on Empirical Methods in Natural Language Processing, pp. 632–642 (2015)

55. Williams, A., Nangia, N., Bowman, S.: A broad-coverage challenge corpus for sentence understanding through inference. In: Conference of the North American Chapter of the Association for Computational Linguistics: Human Language Technologies 1, pp. 1112–1122 (2018)

56. Baevski, A., Zhou, Y., Mohamed, A., Auli, M.: wav2vec 2.0: a framework for self-supervised learning of speech representations (2020)
57. Hsu, W.-N., Bolte, B., Hunert Tsai, Y.-H., Lakhotia, K., Salakhutdinov, R., Mohamed, A.: HuBERT: self-supervised speech representation learning by masked prediction of hidden units. IEEE/ACM Trans. Audio Speech Lang. Process. **29**, 3451–3460 (2021)
58. S. Jones, K.: IDF term weighting and IR research lessons. J. Docum. **60**(6), 521–523 (2004)
59. Maiya, A.S.: ktrain: a low-code library for augmented machine learning. arXiv:2004.10703 (2020)
60. Chen, T., Sun, Y., Shi, Y., Hong, L.: On sampling strategies for neural network based collaborative filtering. In: ACM SIGKDD International Conference on Knowledge Discovery and Data Mining, pp. 767–776 (2017)
61. Chen, T., Kornblith, S., Norouzi, M., Hinton, G.: A simple framework for contrastive learning of visual representations. In: International Conference on Machine Learning (ICML), pp. 1597–1607 (2020)
62. Agirre, E., Cer, D., Diab, M., Gonzalez-Agirre, A.: SemEval-2012 task 6: a pilot on semantic textual similarity. In *SEM 2012: The First Joint Conference on Lexical and Computational Semantics - Volume 1: Proceedings of the main conference and the shared task, and Volume 2: Proceedings of the Sixth International Workshop on Semantic Evaluation (SemEval 2012), pp. 385–393 (2012)
63. Agirre, E., Cer, D., Diab, M., Gonzalez-Agirre, A., Guo, W.: *SEM 2013 shared task: Semantic textual similarity. In: Second Joint Conference on Lexical and Computational Semantics (*SEM), volume 1: Proceedings of the Main Conference and the Shared Task: Semantic Textual Similarity, pp. 32–43 (2013)
64. Agirre, E.. et al.: SemEval-2014 task 10: Multilingual semantic textual similarity. In: Proceedings of the 8th International Workshop on Semantic Evaluation (SemEval 2014), pp. 81–91 (2014)
65. Agirre, E., et al.: SemEval-2015 task 2: semantic textual similarity, English, Spanish and pilot on interpretability. In: Proceedings of the 9th International Workshop on Semantic Evaluation (SemEval 2015), pp. 252–263 (2015)
66. Agirre, E., et al.: SemEval-2016 task 1: semantic textual similarity, monolingual and cross-lingual evaluation. In: Proceedings of the 10th International Workshop on Semantic Evaluation (SemEval-2016), pp. 497–511. Association for Computational Linguistics (2016)
67. Cer, D., Diab, M., Agirre, E., Lopez-Gazpio, I., Specia, L.: SemEval-2017 task 1: semantic textual similarity multilingual and crosslingual focused evaluation. In: Proceedings of the 11th International Workshop on Semantic Evaluation (SemEval-2017), pp. 1–14 (2017)
68. Marelli, M., Menini, S., Baroni, M., Bentivogli, L., Bernardi, R., Zamparelli, R.: A SICK cure for the evaluation of compositional distributional semantic models. In: International Conference on Language Resources and Evaluation (LREC), pp. 216–223 (2014)
69. Reimers, N., Beyer, P., Gurevych, I.: Task-oriented intrinsic evaluation of semantic textual similarity. In: International Conference on Computational Linguistics (COLING), pp. 87–96

Light-Weight CNN-Attention Based Architecture Trained with a Hybrid Objective Function for EMG-Based Human Machine Interfaces

Soheil Zabihi[1], Elahe Rahimian[2], Amir Asif[1], Svetlana Yanushkevich[3], and Arash Mohammadi[1,2(✉)]

[1] Electrical and Computer Engineering, Concordia University, Montreal, QC, Canada
arash.mohammadi@concordia.ca
[2] Concordia Institute for Information System Engineering, Concordia University, Montreal, QC, Canada
[3] Biometric Technologies Laboratory, Department of Electrical and Computer Engineering, University of Calgary, Calgary, AB, Canada

Abstract. The presented research focuses on Hand Gesture Recognition (HGR) utilizing Surface-Electromyogram (sEMG) signals. This is due to its unique potential for decoding wearable data to interpret human intent for immersion in Mixed Reality (MR) environments. The existing solutions so far rely on complicated and heavy-weighted Deep Neural Networks (DNNs), which have restricted practical application in low-power and resource-constrained wearable systems. In this work, we propose a light-weight hybrid architecture (HDCAM) based on Convolutional Neural Network (CNN) and attention mechanism to effectively extract local and global representations of the input. The proposed HDCAM model with $58,441$ parameters reached a new state-of-the-art (SOTA) performance with 83.54% and 82.86% accuracy on window sizes of $300\,\mathrm{ms}$ and $200\,\mathrm{ms}$ for classifying 17 hand gestures. The number of parameters to train the proposed HDCAM architecture is $18.87\times$ less than its previous SOTA counterpart. Furthermore, the model is trained based on a hybrid loss function consisting of two-fold: (i) Cross Entropy (CE) loss which focuses on identifying the helpful features to perform the classification objective, and (ii) Supervised Contrastive (SC) loss which assists to learn more robust and generic features by minimizing the ratio of intra-class to inter-class similarity.

Keywords: Attention Mechanism · Biological Signal Processing (BSP) · Mixed Reality (MR) · surface Electromyogram · Hand Gesture Recognition

---

This Project was partially supported by Department of National Defence's Innovation for Defence Excellence & Security (IDEaS), Canada.

M. Gavrilova et al. (Eds.): *Transactions on Computational Science XL*, LNCS 13850, pp. 48–65, 2023.
https://doi.org/10.1007/978-3-662-67868-8_4

1   Introduction

Surface Electromyogram (sEMG)-based Hand Gesture Recognition (HGR) is regarded as a promising approach for a wide range of applications, including myoelectric control prosthesis [1–4], virtual reality technologies [5,6], Human Computer Interactions (HCI) [7], and rehabilitative gaming systems [8]. sEMG signals contain electrical activities of the muscle fibers that can be employed to decode hand gestures and thereby enhance immersive Human Machine Interaction (HMI) wearable systems for immersion in Mixed Reality (MR) environments [9,13]. Consequently, there has been a surge of interest in the development of Deep Neural Networks (DNNs) and Machine Learning (ML) models to identify hand gestures using sEMG signals. Generally speaking, sEMG datasets can be collected based on "sparse multichannel sEMG" or "High-Density sEMG (HD-sEMG)". The latter records electrical activity of muscles by two-dimensional arrays of closely-spaced electrodes, extracting both temporal and spatial changes of muscle action potentials. The advantages of this technique include the ability to obtain a large amount of data and more robustness to electrode changes. Despite advantages of HD-sEMG, its utilization leads to structural complexity [14,15], while adoption of sparse multichannel sEMG signals requires fewer electrodes making it the common modality of choice for incorporation into wearable devices. Therefore, development of DNN models based on sparse sEMG signals has gained significant recent importance. However, more efforts are needed to bridge the gap between academic research and clinical solutions in this area [9].

Despite extensive research in this area and the fact that academic researchers achieve high classification accuracy in laboratory conditions, there is still a gap between academic research in sEMG pattern recognition and commercialized solutions [9]. For instance, one of the main obstacles in current prosthesis devices is the lack of feedback provided to the user regarding the prosthesis's position or the forces being applied. This can make the control process difficult and less precise for the user, leading to less natural and less efficient interaction with the device. To develop a user-friendly and reliable prosthesis control, providing feedback is crucial [9–11]. Moreover, there are challenges related to the wearability and portability of the sEMG-based systems, as well as the ease of use, and the robustness against the variations in muscle activation patterns, which may affect the performance of the systems [12]. Academic researches often focus on developing advanced and sophisticated algorithms to improve the performance of sEMG-based prosthesis control, but these methods may be too complex or too expensive to be practical for industrial use from the time and computation perspective [1–4]. All these factors contribute to the existing gap between academic research in sEMG pattern recognition and commercialized solutions, and further research and development are needed to overcome these limitations and improve the performance and usability of sEMG-based systems for practical applications. In this context, the primary goal of this study is to reduce the gap by developing DNN-based models that not only have high recognition accuracy but also have minimal processing complexity, allowing them to be embedded in low-power devices such as wearable controllers [1,16]. Furthermore, the designed

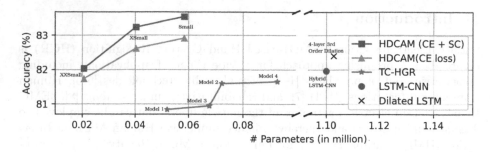

Fig. 1. Comparing different variants of the proposed HDCAM model with SOTA designs for an input window size of 300 ms. The x-axis shows the number of parameters and the y-axis displays the classification accuracy on the Ninapro DB2 dataset. HDCAM shows a better compute versus accuracy trade-off compared to recent approaches. The square-blue plot shows HDCAM trained with Cross Entropy (CE) and Supervised Contrastive (SC) losses, whereas all other models are trained with only CE loss. (Color figure online)

DNN-based models should be based on the minimum number of electrodes while estimating the desired gestures within an acceptable delay time [9,17]. Consequently, in this paper, we develop the novel **Hierarchical Depth-wise Convolution along with the Attention Mechanism (HDCAM)** model for HGR based on sparse sEMG signals to fill this gap by meeting criteria such as improving the accuracy and reducing the number of parameters. The HDCAM is developed based on the Ninapro [18,19] database, which is one of the most well-known sparse multi-channel sEMG benchmark datasets.

Using Convolutional Neural Networks (CNN) [20–23] is a common approach for hand movement classification, where sEMG signals are first converted into images and then used as input for CNN-based architectures. However, the nature of sEMG signals is sequential, and CNN architectures only take into account the spatial features of the sEMG signals. Therefore, in recent literature [15,16,24], authors proposed using recurrent-based architectures such as Long Short Term Memory (LSTM) networks to exploit the temporal features of sEMG signals. On the other hand, it is suggested in [25–27] to use hybrid models (CNN-LSTM architecture) instead of using a single model to capture the temporal and spatial characteristics of sEMG signals. Although recent academic researchers are improving the performance by using Recurrent Neural Networks (RNNs) or hybrid architectures, the sequence modeling with recurrent-based architectures has several drawbacks such as consuming high memory, lack of parallelism, and lack of stable gradient during the training [4,28]. It is demonstrated [29] that sequence modeling using RNN-based models does not always outperform CNN-based designs. Specifically, CNN architectures have several advantages over RNNs such as lower memory requirements and faster training if designed properly [29]. Therefore, in the recent literature [4,28,30,31], the authors took advantage of 1-D Convolutions developed based on the dilated causal convolutions, where the sequence of sEMG signals can be processed as a whole with lower memory requirement during the training compared to RNNs. Convolution operation in CNNs, however, has two main limitations, i.e., (i) it has a local receptive field, which makes it incapable of modeling global context, and; (ii) their learned weights remain stationary at inference time, therefore, they cannot adapt to changes in input. Attention mechanism [32] can mitigate both of these problems. Consequently, the authors in the recent research papers [1,2,33–35] used the attention mechanism combined with CNNs and/or RNNs to improve the performance of sEMG-based HGR. The attention mechanism's major disadvantage is that it is often computationally intensive. Therefore, a carefully engineered design is required to make attention-based models computationally viable, especially for low-power devices.

In this paper, we develop the HDCAM architecture by effectively combining the complementary advantages of CNNs and the attention mechanisms. Our proposed architecture shows a favorable improvement in terms of parameter reduction and accuracy compared to the state-of-the-art (SOTA) methods for sparse multichannel sEMG-based hand gesture recognition (see Fig. 1). The contributions of the HDCAM architecture can be summarized as follows:

- Efficiently combining advantages of Attention- and CNN-based models and reducing the number of parameters (i.e., computational burden).
- Efficiently extracting local and global representations of the sEMG sequence by coupling convolution and attention-based encoders.
- Integration of Depth-wise convolution ($DwConv$) a hierarchical structure in the proposed Hierarchical Depth-wise Convolution ($HDConv$) encoder, which not only extracts a multi-scale local representation but increases the receptive field in a single block.

The small version of the proposed HDCAM with $58,441$ parameters achieves new SOTA 83.54% top-1 classification accuracy on Ninapro DB2 dataset with $18.87\times$ less number of parameters compared to the previous SOTA approach [16].

2   The Proposed HDCAM Architecture

The primary objective of this study is to build a light-weight hybrid architecture that successfully combines advantages of Attention- and CNN-based models for low-powered devices. In what follows, we first briefly describe the preprocessing steps for sEMG data preparation, and then the HDCAM architecture is explained in detail. Finally, the training objectives are described.

2.1   Preprocessing Step

For the pre-processing step, we employed the first order 1 Hz low-pass Butterworth filter to smooth the raw sEMG signals as described in the recent literature [19,20,36,37]. Moreover, we normalized and scaled the sEMG signals by the μ-law technique introduced in [3]. The μ-law is applied to preserve the scale of sensors with larger values while amplifying the range of sensors with smaller values. It is demonstrated that normalizing sEMG signals using the μ technique

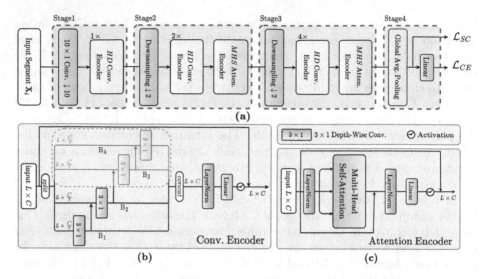

Fig. 2. The proposed architecture: **(a)** The overall architecture of proposed HDCAM model. At the stage 4, the output representations of the Global Average Pooling (GAP) layer are passed to Supervised Contrastive Loss ($\mathcal{L}_{SC}$), and the output logits of the Linear layer are used in Cross Entropy Loss ($\mathcal{L}_{CE}$). **(b)** The *HDConvEncoder* uses Hierarchical Depth-wise Convolution for multi-scale temporal feature mixing followed by a point-wise convolution, i.e. Linear layer, for channel mixing. To expand the receptive field in the deeper layers, the number of active branches (B_i) is increased from Stage 1 to Stage 3. **(c)** The design of the *MHSAttenEncoder* is illustrated, which consists of a Multi-Head Self-Attention (MHA) mechanism to encode the global representation of the input feature maps.

achieves better performance than using the Min-Max normalization [3]. The formulation of the *μ-law* normalization is as follows

$$F(x_t) = \text{sign}(x_t) \frac{\ln\left(1 + \mu |x_t|\right)}{\ln\left(1 + \mu\right)} \tag{1}$$

where the time is denoted by the variable t, the sEMG value at time step t is shown by x_t, and the new range is represented by the variable μ which is set to 256. This step completes the pre-processing step for raw sEMG signals. Next, we describe the proposed HDCAM architecture.

2.2   The HDCAM Architecture

In the proposed framework, a sliding window strategy with window of size $W \in \{150, 200, 250, 300\,\text{ms}\}$ is adapted to use the multi-variate temporal sEMG information, resulting in dataset $\mathcal{D} = \{(\boldsymbol{X}_i, \text{y}_i)\}_{i=1}^{N}$. More specifically, $\text{y}_i \in \mathbb{R}$ is the label assigned to the ith segmented sequence $X_i \in \mathbb{R}^{L \times C}$. Here, L is the length of the segmented sequential input corresponding to the number of samples obtained at a frequency of 2 kHz for a window of size W, and C denotes the number of channels in the input segment corresponding to the number of input features/sensors. As illustrated in Fig. 2, the HDCAM framework has a hybrid design based on CNN and the "Multi-Head Self-Attention (MHA) mechanism" to reap the advantages of both methods for designing a light-weight architecture for low-power devices.

Overview of the Architecture. As shown in Fig. 2(a), the overall HDCAM architecture consists of four different stages, the first three for multi-scale feature extraction and the last one for classification. HDCAM is made up of two primary components, namely "Hierarchical Depth-wise Convolution ($HDConv$)" encoder and "Multi-Head Self-Attention ($MHSAtten$)" encoder, where the former and latter aim to model the local and global information in the sequential input, respectively. Formally, for a given segmented sequential input $X_i \in \mathbb{R}^{L \times C}$, HDCAM begins with the Stem layer. More specifically, the Stem layer serves as a patching mechanism for the input X_i which applies a 10×1 strided convolution with stride of size 10 followed by a Layer Normalization (LN) to the input. Patching mechanism helps reduce memory and computation requirements in downstream layers resulting in $L/10 \times C_1$ feature maps. Afterward, local features are extracted using a $HDConv$ encoder. Further processing of the feature maps takes place in the second and third stages, which follow almost the same architectural structure. Both of which start with the downsampling layer followed by consecutive $HDConv$ encoders for *local* feature extraction and ended with $MHSAtten$ block to encode the *global* representations of the input. The downsampling layer consists of an LN followed by a 2×1 strided convolution with stride of size 2, which reduces the sequential feature maps length by half and increases the number channels, resulting $L/20 \times C_2$ and $L/40 \times C_3$ dimensional features for second and third stages, respectively. In the final stage, a Global Average Pooling (GAP) operation is used to reduce the feature maps' dimension followed by a Linear layer for classification. When Supervised Contrastive (SC) loss is adopted to training model the output of the GAP layer, denoted by $z_i \in \mathbb{R}^{C_3}$, is used as the input X_i representation, further discussed in Sect. 2.3. Here, C_k refers to number of channels in k^{th} stage, for $k \in \{1, 2, 3\}$.

HDConv **Encoder.** As shown in Fig. 2(b), the proposed *HDConv* block combines depth-wise convolution with a hierarchical structure to extract local features at multi-scales. The proposed multi-scale feature extractor is inspired by the Res2Net [38] module, which combines features with different resolutions. Different from the Res2Net module, we omitted the first point-wise convolution layer and added a 3×1 depth-wise convolution to the first branch. Also, the number of active branches in the hierarchical convolutional structure is dynamic and varies depending on the stage. In *HDConv* module, input feature maps of shape $L \times C$ is evenly splitted into s subsets/scales, denoted by x_i of shape $L \times C/s$, where $i \in \{1, 2, \ldots, s\}$. Then, 3×1 depth-wise convolution, denoted by $DwConv_i$, is applied on each subset x_i after combining with the previous branch output features, denoted by y_{i-1}. Generally, we can write the output features of each branch y_i as follows

$$y_i = \begin{cases} DwConv_i(x_i) & i = 1 \\ DwConv_i(x_i + y_{i-1}) & 2 \leq i \leq s \end{cases} \tag{2}$$

As shown in Fig. 2(b) and Eq. (2), the hierarchical structure allows each depth-wise convolution $DwConv_i$ receive the information from all previous splits, $\{x_j, j \leq i\}$. The output feature maps of all branches are concatenated and passed through an LN followed by point-wise convolution to enrich the multi-scale local representation, and finally, Gaussian Error Linear Unit (GELU) activation is used for adding non-linearity to the model. For information flow through the network hierarchy, residual connection is used in $HDConv$ encoder. The $HDConv$ encoder can be represented as follows

$$X_{out} = X_{in} + Linear_{GELU}(LN(HDwConv(X_{in}))) \tag{3}$$

where X_{in} and X_{out} are the $HDConv$ input and output feature maps, both of shape $L \times C$, $Linear_{GELU}$ is point-wise convolution followed by GELU non-linearity LN is Layer Normalization, and $HDwConv$ is hierarchical depth-wise convolution operation. Finally, it worth to note that in order to expand the receptive field in the deeper layers of the network, the number of active branches (B_i in Fig. 2(b)) in $HDConvEncoder$ is increased from Stage 1 to Stage 3 for the proposed model.

MHS Atten Encoder. In [32], the authors showed that the attention mechanism allows a model to present global information in a given input sequence. Furthermore, attention-based architectures [1,2,33–35] have shown promising performance in the context of sEMG-based HGR by extracting particular bits of information from the sequential nature of the sEMG signals. However, most of these models are still heavy-weight to be used in resource-constrained devices. Hence, in the proposed HDCAM architecture, we designed a hybrid architecture that combines convolutions and attention mechanism advantages. Specifically, due to spatial inductive biases in convolution operation, the CNN-based encoder ($HDConv$) assists our hybrid model to learn local representations with fewer parameters than solely attention-based models. However, to effectively learn global representations, we also used an attention-based encoder ($MHSAtten$). Since computation in the MHA has quadratic relation to input size, we only used the $MHSAtten$ encoder in the second and third stages of the HDCAM to efficiently encode the global representation, where the length of the sequential feature maps are $1/20$ and $1/40$ of the original input of the network, respectively. The $MHSAtten$ encoder can be represented as follows

$$X_{out} = Linear_{GELU}(LN(X_{in} + MHA(LN(X_{in})))) + X_{in} \tag{4}$$

where X_{in} and X_{out} are the $MHSAtten$ input and output feature maps, both of shape $L \times C$, $Linear_{GELU}$ is point-wise convolution followed by GELU non-linearity LN is Layer Normalization, and MHA is Multi-Head Self-Attention mechanism. In MHA, the input feature maps X_{in} of shape $L \times C$ are passed through a Linear projection to create Queries Q, i.e., a matrix with the same shape as the input feature maps. Then, Queries Q is evenly splitted into h subsets, denoted by q_i of shape $L \times C/h$, where $i \in \{1, 2, \ldots, h\}$ and h is number of the heads. In parallel, the same approach has been applied to construct Keys and Values subsets, i.e., k_i and v_i. Finally, on each head, the attention block measures the pairwise similarity of each q_i and all k_h to assign a weight to each v_h. The entire operation is

$$A_h = Softmax(\frac{q_h k_h^T}{\sqrt{d}})v_h, \tag{5}$$

where $d = C/h$ denotes the dimension of k_h and q_h subsets. Then concatenation of attention feature maps of all heads is projected to get the final attention maps of the MHA mechanism, i.e., $MHA(X_{in}) = Linear(Concat(A_1, A_2, \ldots, A_h))$. This completes the description of the proposed HDCAM architecture, next, we present our results to evaluate its HGR performance.

2.3   Training Objectives

For model training, we employ a hybrid loss that consists of two-fold: **(i)** Cross Entropy (CE) loss which focuses on identifying the helpful features to perform the classification objective, and **(ii)** Supervised Contrastive (SC) loss which assists to learn more robust and generic features by minimizing the ratio of intra-class to inter-class similarity.

Cross Entropy (CE) Loss. To train a classifier by CE loss, the predicted probability of each sample X_i is compared to the actual expected value y_i, and a loss is calculated to penalize model weights θ based on how far the prediction is from the actual expected value. Given a training batch $\mathcal{B} = \{(X_i, y_i)\}_{i=1}^{|\mathcal{B}|}$, the CE is formulated as

$$\mathcal{L}_{CE} = -\frac{1}{|\mathcal{B}|} \sum_{i=1}^{|\mathcal{B}|} y_i \log p_\theta(y_i | X_i) \tag{6}$$

where $p_\theta(y_i | X_i)$ is predicted class probability by the classifier. Although CE loss is the most commonly used objective function to adjust weights of deep classification models, it has several known issues, such as the lack of robustness to noisy labels [39] and the possibility of inefficient margins [40]. Hence to mitigate these limitations inspired by recent works [41, 42], we added SC loss [43] as a regularization term to the conventional CE objective function.

Supervised Contrastive (SC) Loss. The SC loss is intended to increase the similarity between features resulted from positive sets while simultaneously driving away features of the negative sets. Following [43], to form positive and negative sets, we leverage label information. For instance, given a training batch

$\mathcal{B} = \{(\boldsymbol{X}_i, y_i)\}_{i=1}^{|\mathcal{B}|}$, for a sample $\boldsymbol{X}_i$ (i.e. anchor), the anchor set is all samples in the batch except $\boldsymbol{X}_i$, and the positive set is composed of the samples that are in the same class as $\boldsymbol{X}_i$, i.e., samples with the label y_i. Accordingly, the negative set is defined by the samples that are in the anchor set but not in the positive set. To compute the SC loss, we first embed inputs in lower dimension space to get the representations, denoted by $\mathbf{z}_\star$. Then, the SC loss can be computed by

$$\mathcal{L}_{\text{SC}} = -\frac{1}{|\mathcal{B}|} \sum_{\forall i \in \mathcal{I}} \log \frac{\sum_{\forall p \in \mathcal{P}(i)} \exp(\mathbf{z}_i \cdot \mathbf{z}_p)}{\sum_{\forall a \in \mathcal{A}(i)} \exp(\mathbf{z}_i \cdot \mathbf{z}_a)} \tag{7}$$

where $\cdot$ denotes the dot product operation, $\mathcal{I} = \{1, 2, \ldots, |\mathcal{B}|\}$ indicates indices of all samples in the batch, $i \in \mathcal{I}$ is the index of the anchor, $\mathcal{A}(i) \equiv \mathcal{I} \backslash \{i\}$ represents the indices of all batch samples but the anchor, and $\mathcal{P}(i) = \{p \in \mathcal{A}(i) : y_i = y_p\}$ is positive set composed of the indices of samples sharing the same class. In our framework, the output feature maps of the GAP layer in stage 4 are used as the representations $\mathbf{z}_\star$ of the inputs (see Fig. 2(a)).

Hybrid Loss. According to [43], using the conventional SC loss requires two distinct training stages for a classification problem: first, learning the representations using SC loss, and second, training classifier on top of the learned representations with the CE loss. However, SC loss generally demands a relatively high batch size in order to get acceptable and stable performance, while this is not the case for CE loss. Therefore, to take the advantages of both CE and SC losses, we jointly trained the HDCAM with the weighted sum of them as follow

$$\mathcal{L}_{\text{H}} = \mathcal{L}_{\text{CE}} + \lambda \mathcal{L}_{\text{SC}} \tag{8}$$

where λ is a weighting coefficient for balancing the losses. Using weighted sum of losses, we can have end-to-end training and learn more general and robust representation due to the minimization of the intra- to inter-class similarity ratio.

3   Experiments and Results

In this section, the description of the database utilized for the evaluation of the HDCAM architecture is presented first. Then, different conducted experiments and their outcomes are provided.

3.1   Database

The proposed HDCAM is trained and tested using the second Ninapro dataset [19] referred to as the DB2, which is the most commonly used sparse sEMG benchmark. In the DB2 dataset, muscle electrical activities are measured by the Delsys Trigno Wireless EMG system with 12 electrodes at a frequency

Table 1. HDCAM Architecture variants. Description of the models' layers with respect to kernel size, and output channels, repeated n times. We use a hierarchical structure in *HDConvEncoder* to extract multi-scale local features. Also, *MHSAttenEncoder* is used to extract global representations of the feature maps.

	Layer (n)	#Layers	Kernel Size	Output Channels		
				XXSmall	**XSmall**	**Small**
Stage1	Stem	1	10×1	16	24	24
	*HD*Conv Encoder	1	3×1	$16(s = 2)$	$24(s = 3)$	$24(s = 3)$
Stage2	Downsampling	1	2×1	24	32	32
	*HD*Conv Encoder	2	3×1	$24(s = 3)$	$32(s = 4)$	$32(s = 4)$
	*MHS*Atten Encoder	1	–	$24(h = 3)$	$32(h = 4)$	$32(h = 4)$
Stage3	Downsampling	1	2×1	32	48	64
	*HD*Conv Encoder	4	3×1	$32(s = 4)$	$48(s = 4)$	$64(s = 4)$
	*MHS*Atten Encoder	1	–	$32(h = 4)$	$48(h = 4)$	$64(h = 4)$
	Global Avg. Pooling	1	–	32	48	64
	Linear	1	1	17	17	17
	Model Parameters			$20,689$	$40,281$	$58,441$

rate of 2 kHz. The DB2 dataset consists of 50 different hand movements (including rest) recorded from 40 healthy users. Each gesture is repeated six times by each user, where each repetition lasts for 5 s, followed by 3 s of rest. For a fair comparison, as well as following the recommendations of the database [19] and previous literature [1,16,27,33], we considered two repetitions (i.e., 2 and 5) for testing and the rest for training. The DB2 dataset is presented in three sets of exercises (B, C, and D). Following [1,16,27], the focus is on Exercise B which consists of 17 hand movements. Next, the experiments and results to evaluate our proposed HDCAM model are provided.

3.2   Results and Discussions

In this section, a comprehensive set of experiments is conducted to evaluate the performance of the proposed HDCAM architecture. Table 1 represents the sequence of the *HDConv* and *MHSAtten* encoders along with design information of the extra-extra small (XXSmall), extra-small (XSmall), and small (Small) versions of the model. As shown in Table 1, the type, number, and sequence of the component blocks in the overall model architecture (illustrated in Fig. 2) are maintained across all HDCAM architecture variants. The number of output channels in each stage is what distinguishes the XXSmall, XSmall, and Small models from one another. Since the number of active branches (s) and attention heads (h) in the *HDConv* and *MHSAtten* encoders is proportional to the number of output channels of the corresponding stage, we maintained the fundamental rule for all model variants, which requires having at least eight channels per-head/per-branch. Additionally, the maximum allowed number of heads/branches is set to four. All models were trained using the Adam optimizer at a learning rate of 10^{-4}. We trained HDCAM with only using CE loss (Eq. 6), and also with hybrid loss (Eq. 8) in which we empirically set to $\lambda = 0.25$. Furthermore, during training, the number of samples for each class in a batch is set to 32, leading to a balance batch of size 544 samples. In the following sections, we conducted several experiments to evaluate our proposed HDCAM model. It is worth mentioning that in certain experiments, models were exclusively trained using CE loss to exclude the influence of the SC loss on the outcomes.

Impact of Contrastive Loss. Table 2 illustrates the average recognition accuracy of different variants of HDCAM over all subjects with and without SC loss function involvement in model training. It can be observed that the performance of all variants of HDCAM for all window sizes is improved when SC loss was involved in training. For instance, the best performance is archived for Small model with window size of 300 ms in Table 2 which is 82.91%, while this value increased by 0.63% when SC is used along with CE loss. These performance improvements demonstrate the usefulness of the SC loss in improving the quality of the learned representation.

The Model's Dimension. This experiment analyzes the recognition accuracy of the HDCAM by varying the number of channels in each stage, yielding XXSmall, XSmall, and Small models. In this regard, Table 2 shows the results for all variants of the proposed architecture for different window sizes. For the same arrangement of component layers, it can be seen from Tables 1 and 2 that the accuracy of the model is improved by increasing the dimensions of the stages regardless of training with hybrid loss or sole CE loss. More specifically, the dimension of the stage 3 is the only difference between the XSmall and Small architectures, resulting in more informative high-level features in the Small model, which leads to better performance. Comparing XXSmall versus two other variants, the dimension of all stages has reduced leading to lower performance. From Table 1 and 2, it can be observed that there is a trade-off between the complexity of the model and the accuracy.

The Effect of Window Size. As shown in Table 2, sliding window strategy with window of size $W \in \{150, 200, 250, 300 \, \text{ms}\}$ is adapted to evaluate the performance of the HDCAM. It is worth mentioning that W is required to be under 300 ms to have a real-time response in peripheral human machine intelligence systems [17]. Comparing outcomes in each column of Table 2 shows that increasing window size (W) led to better performance for all model variants for both losses. According to this observation, the proposed HDCAM architecture is capable of extracting/utilizing information from longer sequences of inputs. For instance, for XXSmall, XSmall, and Small architectures, increasing W from

Table 2. Accuracy of HDCAM variants trained with hybrid loss ($\lambda = 0.25$) and only CE loss over different window sizes (W).

	Model ID	XXSmall	XSmall	Small	XXSmall	XSmall	Small
	Loss	$\mathcal{L}_H = \mathcal{L}_{CE} + \lambda * \mathcal{L}_{SC}$			$\mathcal{L}_{CE}$		
$W = 150$ ms	Accuracy (%)	80.82	82.01	82.44	80.53	81.51	82.21
	STD (%)	6.6	6.2	6.3	6.9	6.6	6.7
$W = 200$ ms	Accuracy (%)	81.34	82.66	82.86	81.10	81.77	82.28
	STD (%)	6.7	6.7	6.5	6.8	6.8	6.6
$W = 250$ ms	Accuracy (%)	81.73	82.82	83.13	81.26	82.17	82.57
	STD (%)	6.8	6.6	6.6	6.8	6.7	6.6
$W = 300$ ms	Accuracy (%)	82.03	83.23	83.54	81.73	82.61	82.91
	STD (%)	6.6	6.8	6.3	6.7	6.6	6.5

150 ms to 300 ms resulted in accuracy improvements of 1.21%, 1.22%, and 1.1% when trained with hybrid loss, respectively, These values are changed to 1.2%, 1.1%, and 0.7% when models are trained withe sole CE loss. Although the number of parameters for a specific version of the model does not change for different W, larger W leads to longer sequential feature maps in the second and third stages, which leads to more memory requirements in the attention mechanism. We would like to emphasize that our proposed hybrid architecture is still far superior to the sole attention-based approach since the sequence lengths in the second and third stages are significantly decreased, as previously noted.

Comparison with State-of-the-Art (SOTA). All of the SOTA methods mentioned in Table 3 are trained by the CE loss. In Table 3, HDCAM is compared with recent SOTA recurrent (Dilated LSTM) [16], convolutional (CNN), hybrid LSTM-CNN [27], and hybrid attention-CNN [1] models on Ninapro DB2 dataset [19]. Overall, our model demonstrates better accuracy versus the number of parameters compared to other methods regardless of training objective function. As shown in Table 3, for the window size of 200 ms, all variants of the proposed model outperform other SOTA approaches with and without SC loss. For instance, our XXSmall model has 53.3 times less parameter than Dilated LSTM, but obtains a 2.1% (2.34%) gain in the top-1 accuracy when trained with sole CE loss (hybrid loss). Compared to the best performing TC-HGR model (Model 4), our XXSmall and Small models trained with CE loss improve the accuracy for 0.38% and 1.56% with 4.59 and 1.62 times less number of parameters, respectively. Moreover, as shown in Table 3, for the window size of 300 ms, XSmall and Small variants of HDCAM, trained with CE loss, obtain 82.61% and 82.91% top-1 accuracy respectively, both surpassing all previous SOTA methods with fewer number of parameters (see Fig. 1). Dilation-based LSTM [16], as the previous SOTA model on DB2 dataset, reached 82.4% top-1 accuracy with $1,102,801$ number of parameters, while our XSmall model attains better accuracy (82.61%)

with only $40,281$ parameters, i.e., 27.38 times fewer. It is worth noting that our Small model achieved 82.91% top-1 accuracy with $58,441$ parameters trained with CE loss. Small model reaches a new SOTA performance even with sole CE loss training that demonstrates the effectiveness and the generalization of our design.

In Table 4, the computation reduction of the proposed model is investigated. More specifically, Table 4 provides average inference time for different variants of the HDCAM and TC-HGR. It is important to note that the processing time can vary depending on the hardware used. In this study, we utilized a GeForce GTX 1080 Ti Graphics Cards to obtain the average inference computation time of each model per input sample. The results, as shown in Table 4, demonstrate that the inference computation time of all variants of the HDCAM model are smaller in comparison to computation times the TC-HGR model variants, while improving performance as shown in Table 3.

Effectiveness of the Multi-scales Local Representation. To extract multi-scale local features, we integrated depth-wise convolution ($DwConv$) with a hierarchical structure in the proposed $HDConv$ encoder. The hierarchical structure besides the multi-scale feature extraction increases the receptive field in a single block. As shown in Table 5, replacing the *"hierarchical"* $DwConv$ structure in $HDConv$ with a standard $DwConv$ layer degrades the accuracy in all variants of HDCAM, indicating its usefulness in our design. As an example, the top-1 accuracy of the Small model decreased by 0.56% in its non-hierarchical variant.

Importance of Using *MHS* Atten Encoders. To examine the importance of *MHSAtten* encoder, we conducted two ablation studies using this encoder at different stages of the network for $W = 300$ ms. In Table 6, we kept the total number of $HDConv$ and *MHSAtten* encoders fixed to $[1,3,5]$ for experiment 1 to 4. While in experiment 5 to 8, the number of $HDConv$ encoder is set to $[1,2,4]$ for all stages, and *MHSAtten* encoder is progressively added to the end of stages. According to both experiments, adding the *MHSAtten* encoder gradually in the last two stages increases accuracy and the number of parameters. In addition, adding a global *MHSAtten* encoder to the first stage is not beneficial since the features in this stage are not mature enough. When at least one *MHSAtten* encoder is used in the network architecture, the best trade-off between accuracy and the number of parameters obtained for the Small model in both experiments, the highlighted rows in Table 6. Furthermore, we conducted another experiment to investigate the impact of using the *MHSAtten* encoder at the beginning (after downsampling) versus end of each stage on the HDCAM architecture. As shown in Table 7, better performance is achieved by using the *MHSAtten* encoder as the final block of the stages. In other words, it is more beneficial to encode global representations after extracting local representations rather than the other way around.

Table 3. Comparing the performance of the proposed HDCAM models with state-of-the-art (SOTA) models on Ninapro DB2 dataset [19]. Our model in the number of parameters and accuracy outperforms the SOTA models.

Model	Model's Variant	$W = 200$ ms		$W = 300$ ms	
		Parameters↓	Accuracy↑ (%)	Parameters↓	Accuracy↑ (%)
Dilated LSTM [16]	4-layer 3rd Order Dilation	1,102,801	79.0	1,102,801	82.4
	4-layer 3rd Order Dilation (pure LSTM)	–	–	466,944	79.7
LSTM-CNN [27]	CNN	–	–	≈ 1.4M	77.30
	Hybrid LSTM-CNN	–	–	≈ 1.1M	81.96
TC-HGR [1]	Model 1	49,186	80.29	52,066	80.84
	Model 2	68,445	80.63	72,285	81.59
	Model 3	69,076	80.51	67,651	80.95
	Model 4	94,965	80.72	92,945	81.65
HDCAM (CE loss)	XXSmall	**20,686**	81.10	**20,686**	81.73
	XSmall	40,281	81.77	40,281	82.61
	Small	58,441	**82.28**	58,441	**82.91**
HDCAM (Hybrid loss)	XXSmall	**20,686**	81.34	**20,686**	82.03
	XSmall	40,281	82.66	40,281	83.23
	Small	58,441	**82.86**	58,441	**83.54**

Table 4. Comparing average process time of different variants of HDCAM and TC-HGR for hand gesture recognition on window size of 200 ms. The process times are reported in millisecond (ms).

Model	Model's Variant	Process Time Per-Sample ↓ (ms)
TC-HGR [1]	Model 1	3.094
	Model 2	3.305
	Model 3	3.425
	Model 4	3.540
HDCAM	XXSmall	2.317
	XSmall	2.626
	Small	2.859

Table 5. Evaluating the effectiveness of the multi-scales local representation extraction in the *HDConv* encoder for window size 300 ms. For Hierarchical models, the scale values (s) for each stage are provided in Table 1. For Non-hierarchical models, s is equal to 1 at all stages.

HDConvEncoder	Accuracy (%)		
	XXSmall	XSmall	Small
Hierarchical structure	81.73	82.61	82.91
Non-hierarchical structure	81.27	82.21	82.35

Table 6. Evaluating the impact of using *MHSAtten* encoder at a different stage of the network for the window size of 300 ms. The listed values show the number of the corresponding encoder in stages 1 to 3 in order. Highlighted rows indicate the Small model.

ID: Model Configuration	Accuracy↑ (%)	Parameters↓
1 : $HDConv = [1, 3, 5], MHSAtten = [0, 0, 0]$	81.56	37, 673
2 : $HDConv = [1, 3, 4], MHSAtten = [0, 0, 1]$	82.45	54, 249
3 : $HDConv = [1, 2, 4], MHSAtten = [0, 1, 1]$	82.91	58, 441
4 : $HDConv = [0, 2, 4], MHSAtten = [1, 1, 1]$	82.26	60, 817
5 : $HDConv = [1, 2, 4], MHSAtten = [0, 0, 0]$	81.93	31, 785
6 : $HDConv = [1, 2, 4], MHSAtten = [0, 0, 1]$	82.55	52, 969
7 : $HDConv = [1, 2, 4], MHSAtten = [0, 1, 1]$	82.91	58, 441
8 : $HDConv = [1, 2, 4], MHSAtten = [1, 1, 1]$	82.48	61, 585

Table 7. Evaluating the impact of using *MHSAtten* encoder at the beginning vs. end of each stage for the window size of 300 ms.

MHSAttenEncoder	Accuracy (%)		
	XXSmall	XSmall	Small
Fist block of Stage ($MHSAtten = [0, 1, 1]$)	81.31	82.37	82.70
Latest block of Stage ($MHSAtten = [0, 1, 1]$)	81.73	82.61	82.91

4   Conclusion

In this paper, a novel resource-efficient architecture, referred to as the HDCAM, is developed for HGR from sparse multichannel sEMG signals. In comparison to SOTA methods, HDCAM is more effective in terms of both parameters and performance. Its light-weight design is a key step toward incorporating DNN models into wearable for immersive HMI. HDCAM is developed by effectively combining the advantages of Attention-based and CNN-based models for low-powered devices. Specifically, HDCAM is empowered with convolution and attention-based encoders, namely *HDConv* and *MHSAtten*, to efficiently extract local and global representations of the input sEMG sequence. We showed that by proper design of convolution-based architectures, we not only can extract a multi-scale local representation but also can increase the receptive field in a single block. A path for future study is to address the issue of misplacement or relocation of sEMG electrodes/sensors, which is an active field of research. This may be seen as both a constraint for the current study and a promising area for future investigation, which is the subject of our continuing research.

Conflicts of Interests/Competing Interests. The authors declare no competing interests.

References

1. Rahimian, E., Zabihi, S., Asif, A., Farina, D., Atashzar, S.F., Mohammadi, A.: Hand gesture recognition using temporal convolutions and attention mechanism. In: IEEE International Conference on Acoustics, Speech and Signal Processing (ICASSP), pp. 1196–1200 (2022)
2. Rahimian, E., Zabihi, S., Asif, A., Atashzar, S.F., Mohammadi, A.: Few-shot learning for decoding surface electromyography for hand gesture recognition. In: IEEE International Conference on Acoustics, Speech and Signal Processing (ICASSP), pp. 1300–1304 (2021)
3. Rahimian, E., Zabihi, S., Atashzar, F., Asif, A., Mohammadi, A.: XceptionTime: independent time-window XceptionTime architecture for hand gesture classification. In: International Conference on Acoustics, Speech, and Signal Processing (ICASSP), pp. 1304–1308 (2020)
4. Tsinganos, P., Cornelis, B., Cornelis, J., Jansen, B., Skodras, A.: Improved gesture recognition based on sEMG signals and TCN. In: International Conference on Acoustics, Speech, and Signal Processing (ICASSP), pp. 1169–1173 (2019)
5. Ovur, S.E., et al.: A novel autonomous learning framework to enhance sEMG-based hand gesture recognition using depth information. Biomed. Sig. Process. Control **66**, 102444 (2021)
6. Toledo-Peral, C.L., et al.: Virtual/augmented reality for rehabilitation applications using electromyography as control/biofeedback: systematic literature review. Electronics. Electronics **14**(11), 2271 (2022)
7. Guo, L., Lu, Z., Yao, L.: Human-machine interaction sensing technology based on hand gesture recognition: a review. IEEE Trans. Hum.-Mach. Syst. (2021)
8. Mongardi, A., et al.: Hand gestures recognition for human-machine interfaces: a low-power bio-inspired armband. IEEE Trans. Biomed. Circuits Syst. (2022)
9. Farina, D., et al.: The extraction of neural information from the surface EMG for the control of upper-limb prostheses: emerging avenues and challenges. Trans. Neural Syst. Rehabil. Eng. **22**(4), 797–809 (2014)
10. Castellini, C., et al.: Proceedings of the first workshop on peripheral machine interfaces: going beyond traditional surface electromyography. Front. Neurorobot. **8**, 22 (2014)
11. Dhillon, G.S., Horch, K.W.: Direct neural sensory feedback and control of a prosthetic arm. IEEE Trans. Neural Syst. Rehabil. Eng. **13**(4), 468–472 (2005)
12. Milosevic, B., Benatti, S., Farella, E.: Design challenges for wearable EMG applications. In: Design, Automation and Test in Europe Conference and Exhibition, pp. 1432–1437 (2017)
13. Han, B., Schotten, H.D.: Multi-sensory HMI for human-centric industrial digital twins: a 6G vision of future industry. In: IEEE Symposium on Computers and Communications (ISCC), pp. 1–7 (2022)
14. Qu, Y., Shang, H., Li, J., Teng, S.: Reduce surface electromyography channels for gesture recognition by multitask sparse representation and minimum redundancy maximum relevance. J. Healthc. Eng. (2021)
15. Toro-Ossaba, A., et al.: LSTM recurrent neural network for hand gesture recognition using EMG signals. Appl. Sci. **12**(9), 9700 (2022)
16. Sun, T., Hu, Q., Gulati, P., Atashzar, S.F.: Temporal dilation of deep LSTM for agile decoding of sEMG: application in prediction of upper-limb motor intention in NeuroRobotics. IEEE Robot. Autom. Lett. (2021)

17. Hudgins, B., Parker, P., Scott, R.N.: A new strategy for multifunction myoelectric control. IEEE Trans. Biomed. Eng. **40**(1), 82–94 (1993)
18. Atzori, M., et al.: A benchmark database for myoelectric movement classification. Trans. Neural Syst. Rehabil. Eng. (2013)
19. Atzori, M., et al.: Electromyography data for non-invasive naturally-controlled robotic hand prostheses. Sci. Data **1**(1), 1–13 (2014)
20. Geng, W., et al.: Gesture recognition by instantaneous surface EMG images. Sci. Rep. **6**, 36571 (2016)
21. Wei, W., Wong, Y., Du, Y., Hu, Y., Kankanhalli, M., Geng, W.: A multi-stream convolutional neural network for sEMG-based gesture recognition in muscle-computer interface. Pattern Recogn. Lett. (2017)
22. Ding, Z., et al.: sEMG-based gesture recognition with convolution neural networks. Sustainability **10**(6), 1865 (2018)
23. Wei, W., et al.: Surface electromyography-based gesture recognition by multi-view deep learning. IEEE Trans. Biomed. Eng. **66**(10), 2964–2973 (2019)
24. Simao, M., Neto, P., Gibaru, O.: EMG-based online classification of gestures with recurrent neural networks. Pattern Recogn. Lett. 45–51 (2019)
25. Rahimian, E., Zabihi, S., Atashzar, S.F., Asif, A., Mohammadi, A.: Surface EMG-based hand gesture recognition via hybrid and dilated deep neural network architectures for neurorobotic prostheses. J. Med. Robot. Res. 1–12 (2020)
26. Karnam, N.K., Dubey, S.R., Turlapaty, A.C., Gokaraju, B.: EMGHandNet: a hybrid CNN and Bi-LSTM architecture for hand activity classification using surface EMG signals. Biocybern. Biomed. Eng. **42**(1), 325–340 (2022)
27. Gulati, P., Hu, Q., Atashzar, S.F.: Toward deep generalization of peripheral EMG-based human-robot interfacing: a hybrid explainable solution for neurorobotic systems. IEEE Robot. Autom. Lett. **6**(2), 2650–2657 (2021)
28. Rahimian, E., Zabihi, S., Atashzar, S.F., Asif, A., Mohammadi, A.: Semg-based Hand gesture recognition via dilated convolutional neural networks. In: Global Conference on Signal and Information Processing, GlobalSIP (2019)
29. Bai, S., Kolter, J.Z., Koltun, V.: An empirical evaluation of generic convolutional and recurrent networks for sequence modeling. arXiv preprint arXiv:1803.01271 (2018)
30. Tsinganos, P., Jansen, B., Cornelis, J., Skodras, A.: Real-time analysis of hand gesture recognition with temporal convolutional networks. Sensors **22**(5), 1694 (2022)
31. Rahimian, E., Zabihi, S., Asif, A., Atashzar, S.F., Mohammadi, A.: Trustworthy adaptation with few-shot learning for hand gesture recognition. In: IEEE International Conference on Autonomous Systems (ICAS), pp. 1–5 (2021)
32. Vaswani, A., et al.: Attention is all you need. arXiv preprint arXiv:1706.03762 (2017)
33. Rahimian, E., Zabihi, S., Asif, A., Farina, D., Atashzar, S.F., Mohammadi, A.: FS-HGR: few-shot learning for hand gesture recognition via ElectroMyography. IEEE Trans. Neural Syst. Rehabil. Eng. (2021)
34. Wang, S., et al.: Improved multi-stream convolutional block attention module for sEMG-based gesture recognition. Front. Bioengineering Biotechnol. **10** (2022)
35. Hu, Y., et al.: A novel attention-based hybrid CNN-RNN architecture for sEMG-based gesture recognition. PLoS ONE **13**(10), e0206049 (2018)
36. Wei, W., et al.: A multi-stream convolutional neural network for sEMG-based gesture recognition in muscle-computer interface. Pattern Recogn. Lett. **119**, 131–138 (2019)

37. Atzori, M., Cognolato, M., Müller, H.: Deep learning with convolutional neural networks applied to electromyography data: a resource for the classification of movements for prosthetic hands. Front. Neurorobot. **10**, 9 (2016)

38. Gao, S.H., Cheng, M.M., Zhao, K., Zhang, X.Y., Yang, M.H., Torr, P.: Res2net: a new multi-scale backbone architecture. IEEE Trans. Pattern Anal. Mach. Intell. **43**(2), 652–662 (2019)

39. Zhang, Z., Sabuncu, M.: Generalized cross entropy loss for training deep neural networks with noisy labels. In: Advances in Neural Information Processing Systems, pp. 8778–8788 (2018)

40. Liu, W., Wen, Y., Yu, Z., Yang, M.: Large-margin softmax loss for convolutional neural networks. In: International Conference on Machine Learning (ICML), vol. 2, p. 7 (2016)

41. Huang, G., Ma, F.: ConCAD: contrastive learning-based cross attention for sleep apnea detection. In: Joint European Conference on Machine Learning and Knowledge Discovery in Databases, pp. 68–84 (2021)

42. Jeon, S., Hong, K., Lee, P., Lee, J., Byun, H.: Feature stylization and domain-aware contrastive learning for domain generalization. In: Proceedings of the 29th ACM International Conference on Multimedia, pp. 22–31 (2021)

43. Khosla, P., et al.: Supervised contrastive learning. Adv. Neural. Inf. Process. Syst. **33**, 18661–18673 (2020)

Fairness, Bias and Trust in the Context of Biometric-Enabled Autonomous Decision Support

Kenneth Lai$^{(\boxtimes)}$ (iD), Svetlana N. Yanushkevich (iD), and Vlad Shmerko (iD)

Biometric Technologies Laboratory, Schulich School of Engineering,
University of Calgary, Calgary, Canada
{kelai,syanshk,vshmerko}@ucalgary.ca
https://www.ucalgary.ca/labs/biometric-technologies/home

Abstract. Developing a trustworthy biometric-enabled autonomous and semi-autonomous system involves computing the amount of bias and fairness of the data upon which the decisions are made. In this paper, we evaluate the performance of a system that performs human users' face identification in terms of hit and miss rates. We also assess how the usage of data for training biometric systems can lead to unfairness, especially in the context of evolving biometric traits in times of pandemics. In particular, we contrast the performance of normal "unmasked" facial images with their "masked" counterparts. The SpeakingFace and Thermal-Mask dataset is used for assessment, where the Thermal-Mask dataset consists of synthetically applied masks on both the thermal (IR) and visual (RGB) domains. The comparative experiment assesses how fairness metrics such as demographic parity difference (DPD) and equalized odds difference (EOD) can be applied to both masked and normal facial images across the thermal and visual domains. For age, gender, and ethnicity demographic groups, the DPD approaches 0.10%, 0.01%, and 0.01% when the hit rate and miss rate for facial identification approach 100% and 0%, respectively. For masked-thermal facial identification using a simple 2-block convolutional neural network, we obtain a hit rate of 69.96% whereas an 85.82% hit rate is reported for the unmasked alternative. The corresponding age DPD for such hit rate is 49.06% and 34.02% for their respectively masked-thermal and normal-thermal images. We conclude that datasets with inherent biases influence the fairness of a biometric system. The bias manifests itself across the demographic groups (age, gender, and ethnicity), the spectral domain (IR or RGB), and whether synthetic procedures are applied to generate new images. In particular, the error rates among masked facial images are consistently above the unmasked alternative for both visual and thermal domains.

Keywords: Face Biometrics · Human Identification · Fairness · Convolutional Neural Network · Synthetic Image · Face masks · Thermal Face Image · Demographic Parity Difference

© The Author(s), under exclusive license to Springer-Verlag GmbH, DE, part of Springer Nature 2023
M. Gavrilova et al. (Eds.): *Transactions on Computational Science XL*, LNCS 13850, pp. 66–87, 2023.
https://doi.org/10.1007/978-3-662-67868-8_5

1 Introduction

Face recognition plays an important task in the ever-growing domains of computer vision and artificial intelligence (AI). With the adoption of deep machine learning techniques, facial recognition performance has reached an extraordinary 0.1% rank-one miss-rate on a gallery of 12 million individuals [1]. An emerging problem in this domain is the vulnerability to biases which results in unfair decisions [2]. Fair decisions play a critical role in establishing a trustworthy system, therefore unfair decisions have a negative influence on how trustworthy a machine can be. Due to the data-dependent nature of most contemporary machine learning techniques, existing biases in data can bias the underlying algorithms, and, in some cases, may amplify these biases. A biased AI-based decision may lead to unfair treatment such as scenarios in the hiring process [3]. These growing concerns [4] encourage the development of a "fair" AI system which is critical for the future of AI-based decision-making [5,6].

The creation of a "fair" system is a multi-stage process that is dependent on the understanding of bias, and how the mitigation of bias can lead to fairness [7]. In [8], bias is defined as the "inclination or prejudice of a decision made by an AI system which is for or against one person or group, especially in a way considered to be unfair." In [9], fairness is defined as "the absence of any prejudice or favoritism toward an individual or group based on their inherent or acquired characteristics." Fairness in face recognition [10] is linked to the demographics of the subjects [11]. The two most common biases in face biometrics, gender, and racial bias, can manifest themselves in a dataset [12,13]. This is largely due to the nature of data collection, resulting in over or under-representation of the different demographic groups [14]. A possible solution is sub-sampling the over-represented groups [15], which can also be achieved through synthetic augmentation [16] of the under-represented groups.

In this paper, we evaluate the facial recognition performance and fairness metrics for visual and thermal images. We compare this performance with the performance of the synthetic masked counterpart of the above image, to illustrate how the same analysis can be used to assess the fairness of real and synthetic images. The key interest is how the performance and fairness change when a synthetic mask is applied to a normal unmasked facial image as well as how this behavior is the same/different when applied to thermal images. Through this study, we hope to establish a baseline for evaluating the fairness of any autonomous human-machine system which can then be translated to designing and developing trustworthy systems.

2 Motivational Issues: Biometric-Enabled Teammates

For a human operator to accept and cooperate with a decision support tool, it is fundamental to develop trust in the technology. A way to build trust in a facial recognition system generally requires constant exposure to consistent, reasonable, fair, and highly accurate results. Another aspect of building trust is to look at fairness, specifically how the system behaves towards different demographic groups. The system should perform equally well on all demographic groups.

Fairness is one of the core concepts necessary for human collaboration with robots [17]. The collaboration between humans and autonomous robots using biometric-enabled technologies is a challenging area that involves a wide spectrum of physiological and behavioral biometrics [18]. For example, a robot may observe a human teammate looking for human fatigue and intervene when a certain fatigue level is reached [19] or accommodate human operators experiencing high levels of stress [20]. Stress analysis falls in the area of emotion detection/recognition and is generally addressed by face expression analysis. However, faces can often be partially occluded if the subject wears a face mask or other accessories. Addressing this challenge is one focus of this paper.

For training sophisticated recognition systems, specifically for training robot teammates, a large amount of data is required. In some cases, such data is not readily available, which has led to the development of using synthetic biometric data for developing and training robot teammates [21]. The rise of using synthetic data for training has three main reasons [22]: 1) Authentic data are usually collected under limited conditions or scenarios; 2) Critical, boundary-case scenarios may never occur when the authentic data is being collected, e.g. rare events/scenarios such as impersonation, plastic surgery causing face changes, and aging of biometric traits may not be represented in the collected data; and 3) Privacy issues may prevent the developers from accessing or collecting the data; in such case, sensitive personal data must be replaced by synthetic data.

Investigation of the fairness of biometric-enabled systems that utilize synthetic data for model training is an emerging problem. This paper has a two-fold contribution: 1) the process to assess bias in biometric-enabled systems and 2) the framework for evaluating fairness in any system.

In this paper, we focus on the evaluation of fairness as most modern recognition system, such as deep-learning, has reached highly-accuracy predictions. Currently, it is insufficient to only find the best-performing algorithm, and how these algorithms behave under duress. By evaluating the performance under different levels of bias, we can compute the fairness of such algorithms, which, in turn, can lead to trust.

3   Literature Review

Previous work on risk, trust, and bias is proposed in [23], where Trust is defined as a function of the risk of an event occurring and the probability of that event occurring. In terms of the human-machine relationship, trust plays a crucial role in building trustworthy autonomous systems. A user's trust in the system can be built upon a strong foundation of a 1) reliable, 2) explainable, and 3) fair system.

A reliable system is a system that is consistently performing well, in terms of definition, in [1] reliability is defined as the same as a true positive identification rate (TPIR). In addition to TPIR, other common performance measures used to evaluate performance in recognition algorithms include accuracy, cumulative match characteristic, or error rates such as type I and type II errors [24,25]. In order for a system to be reliable, it must report high performance.

An explainable system is another essential component of establishing trust [26]; however, the recent trend of new machine learning techniques, that are achieving the highest results such as deep learning, often are the least explainable [26]. One difficulty of creating explainable artificial intelligence (XAI) is performance vs. interpretability, specifically the focus on ever-increasing performance which increases the demand for more complex data-driven black-box models. Due to the nature of black-box models, it is often less interpretable than white-box models. It is mentioned in [27] that the better approach is to create inherently interpretable models as opposed to trying to explain black-box machine learning models. The XAI goal is to achieve the following criteria: trustworthiness, causality, transferability, informativeness, confidence, fairness, accessibility, interactivity, and privacy awareness [28].

Fairness, another component to trust, is heavily related to reliability and explainability. In [29], fairness is described as making a decision in the absence of discrimination. Bias is one construct that negatively impacts fairness, specifically for data-driven models, inherent bias in the data (imbalanced data) can result in a difference in performance for different demographic groups. Different sources of biases such as data bias, algorithm bias, and user bias are introduced in [9]. In [30], the demographic effects of the facial recognition algorithm are studied, particularly via the use of performance differentials such as false positive and false negative differentials. Additional metrics that can be used to measure fairness include equalized odds [31], demographic parity [4], and counterfactual fairness [32].

For areas of research in artificial intelligence (AI), fairness is also a growing topic. A study by Qian et al. [33] conducted a study to quantify the impact of software implementation on the fairness and variance of deep learning systems. Fairness conditions used for evaluation include statistical parity, predictive equality, and equality of opportunity. In order to develop fair AI, one needs to have an intricate understanding of bias and how unfairness to certain groups of people can impact trustworthiness in AI technology [34].

One popular focus of AI is facial recognition algorithms, spurred by the era of the COVID-19 pandemic where mask-wearing has greatly impacted face recognition algorithms [35–37]. that One of the biases that deeply influences the performance of facial recognition is racial bias. It is reported in [38] among different ethnic groups (Caucasian, Indian, Asian, and African) Caucasians, on average, have the highest performance across multiple commercial APIs and state-of-the-art algorithms on the Racial Faces-in-the-Wild database. Another system proposed in [39] suggests using a combination of mask augmentation, data resampling, and symmetric-arc-loss to obtain fairer results for masked face recognition. It can be seen that machine learning models often learn the incorrect relationship between age, gender, and ethnicity which is hard to discover. A benchmark for studying bias mitigation is proposed in [40] to alleviate this bias problem.

4   Method

In this paper, we address 2 pillars of trust: reliability in terms of performance and fairness in the dataset. We propose an approach to evaluating the performance and fairness of a facial recognition system that can be applied to both real and synthetic images, in both visual and thermal domains. For the experimental dataset, we choose to use the SpeakingFace [41] and Thermal-Mask [42] datasets that include both the visual and thermal modalities. For performance evaluation, we use performance metrics such as hit and miss rates. In addition, we adopt performance differentials, demographic parity and equalize odds to assess the fairness in a facial recognition system. The proposed facial identification utilizes simple 2-Block and 3-Block convolutional neural networks.

4.1   Dataset

SpeakingFaces Dataset [41]: a large-scale multimodal dataset that combines thermal, visual, and audio data streams. It includes data from 142 subjects, with a gender balance of 68 female and 74 male participants, with ages ranging from 20 to 65 years with an average of 31 years. With approximately 4.6 million images collected in both the visible and thermal spectra, each of the 142 subjects has nine different head positions, and each position with 900 frames was acquired in two trials. Figure 1 shows the thermal and visual images of two different subjects.

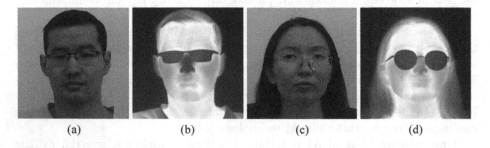

(a)   (b)   (c)   (d)

Fig. 1. Example images from the SpeakingFace Dataset: (a) subject 1 visual, (b) subject 1 thermal, (c) subject 16 visual, (d) subject 16 thermal.

Thermal-Mask dataset [42]: A synthetic mask dataset created using the SpeakingFaces Dataset. This dataset consists of 80 subjects with a total of 84,920 synthetic masked visual and thermal images. The detailed process of applying synthetic masks onto the SpeakingFaces dataset is described in [42]. The general process involves the use of RetinaFace [43] for face detection (pre-trained on the WIDER Face dataset) and HRNet [44] for detecting facial landmarks. The landmarks along with marked mask coordinates are used to fit the mask onto both the visual and thermal facial images. The images in this dataset are cropped and aligned with their SpeakingFaces counterpart and have a pixel resolution of 256×256. Figure 2 shows the thermal and visual mask images of two different subjects.

In this paper, we used 38,222 images from each partition (mask-visual, normal-visual, mask-thermal, normal-thermal) for a total of 152,888 images.

Table 1 shows the demographic group composition of the dataset used. For example, looking at the gender table, we see a composition consisting of 45 male subjects and 35 female subjects. In terms of the number of images per gender group, Table 1 indicates 21,360 images for males and 16,862 images for females.

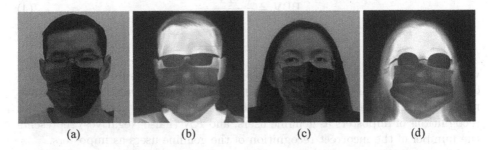

Fig. 2. Example images from the Thermal-Mask Dataset: (a) subject 1 visual, (b) subject 1 thermal, (c) subject 16 visual, (d) subject 16 thermal.

Table 1. Image composition for different demographic groups: Gender, Ethnicity, and Age.

Gender	Male	Female	Ethnicity	A	B	C
# of Images	21360	16862	# of Images	28878	1458	7886
# of Subjects	45	35	# of Subjects	60	3	17

Age	20	21	22	23	24	25	26	27	28	29	30	31
# of Images	2430	4358	1350	5254	1458	2916	1458	3402	1458	2394	969	945
# of Subjects	5	9	3	11	3	6	3	7	3	5	2	2

Age	32	33	34	35	36	37	39	40	41	45	46	57
# of Images	486	972	1918	486	1455	1416	452	486	972	438	263	486
# of Subjects	1	2	4	1	3	3	1	1	2	1	1	1

4.2   Performance Metrics

In this paper, the performance of the machine learning models for face identification is measured in terms of rank-based and threshold-based metrics. Rank-based metrics include: True Positive Rate (TPR, hit rate), False Negative Rate (FNR, miss rate), False Positive Rate (FPR), Positive Predictive Value (PPV), Negative Predictive Value (NPV) defined in Eq. 1, 2, 3, 4, 5. For each of these metrics, rank-1 results are used for positive classification. That is, all rank-1 prediction is treated as the positive prediction. If the rank-1 prediction is the same as the ground-truth then the result is a true positive, otherwise, the result is a false positive.

$$\text{TPR} = \frac{TP}{TP + FN} \tag{1}$$

$$\text{FNR} = \frac{FN}{FN + TP} \tag{2}$$

$$\text{FPR} = \frac{FP}{TP + TN} \tag{3}$$

$$\text{PPV} = \frac{TP}{TP + FP} \tag{4}$$

$$\text{NPV} = \frac{TN}{TN + FN} \tag{5}$$

where TP (True Positives) represents the number of correct subject's recognition of genuine users, TN (True Negatives) represents the number of the correct recognition of imposters, FP (False Positives) represents the number of incorrect recognitions of imposters as genuine users, and FN (False Negatives) represents the number of the incorrect recognition of the genuine users as imposters.

Threshold-based metrics are similar to rank-based metrics except the score-value obtained from each comparison is used to determine a positive or negative prediction as opposed to the rank. In this paper, the threshold-based metrics used are the false positive identification rate (FPIR) and false negative identification rate (FNIR) as defined in Eq. 6 and 7.

$$\text{FPIR(T)} = \frac{FP(T)}{N} \tag{6}$$

$$\text{FNIR(T)} = \frac{FN(T)}{P} \tag{7}$$

where $FP(T)$ is the number of imposters with score values above the threshold T, $FN(T)$ is the number of genuine users below the threshold T, N is the total number of imposter comparisons, and P is the total number of genuine comparisons.

4.3 Fairness Metrics

In this paper, we evaluate the fairness of the facial recognition system using the demographic parity difference [4] and equalized odds difference [31]. Fairness metrics or differentials are an indirect way of measuring the amount of bias the system has, specifically examining how influential the biased cohorts in the training set are. By computing the fairness of a system, a user of such autonomous machines can make an informed decision on whether or not the system is biased towards or against a specific cohort.

Demographic parity, also known as statistical parity, states that for each protected group, their positive rate should be similar. A system satisfies statistical parity if its prediction is statistically independent of the demographic group. This system can be represented as:

$$Pr(\hat{y}|D = a) = Pr(\hat{y}|D = b) = Pr(\hat{y}|D = z) \tag{8}$$

where $\hat{y}$ represents the predictor, D is the demographic group (e.g. gender, ethnicity, etc.), and $a, b, ..., z \in D$ are the classes (e.g. male and female) in the demographic group D.

Demographic parity difference (DPD) is defined as the difference in positive rate between the largest and smallest demographic group. A DPD of 0 represents that all demographic groups have the same positive rate.

$$DPD = Pr(\hat{y}|D = l) - Pr(\hat{y}|D = s) \tag{9}$$

where $l, s \in D$ represents the largest and smallest class in the demographic group D, respectively.

Equalized odds state that the true positive rate and false positive rate across each protected group should be similar. A system satisfies equal opportunity if its prediction is conditionally independent of the protected group. This system can be represented as follows:

$$Pr(\hat{y}|y, D = a) = Pr(\hat{y}|y, D = z) \tag{10}$$

where $\hat{y}$ represents the predictor, y represents conditionally positive outcome, and $a, b, ..., z \in D$ are the classes in protected group D.

Equalized odds difference (EOD) is defined as the larger of the two: true positive rate difference (TPD) and false positive rate difference (FPD). The TPD is defined as the difference in true positive rate between the largest and smallest demographic group. Similarly, the FPD is the difference in false positive rate between the largest and smallest demographic group. The EOD value of 0 represents that all demographic groups have the same true positive, true negative, false positive, and false negative rates.

$$TPD = Pr(\hat{y}|y, D = l) - Pr(\hat{y}|y, D = s) \tag{11}$$

$$FPD = Pr(\hat{y}|y', D = l) - Pr(\hat{y}|y', D = s) \tag{12}$$

$$EOD = Max(TPD, FPD) \tag{13}$$

where y' represents a conditionally negative outcome and $l, s \in D$ represents the largest and smallest class in the demographic group D, respectively.

Performance differentials is the performance difference between the same performance metrics. In this paper, we evaluate the false negative and false positive differentials in terms of rank-based as well as threshold-based approach. These differentials are analyzed in the form of cumulative match characteristic and error tradeoff curves.

Cumulative match characteristic (CMC) curves are an approach used for measuring the facial identification performance [45]. CMC curves are created as a function of rank where rank is the number of top R classes to accept as genuine. For example, a rank-1 result returns only the top ranking class while a rank-3 result returns the top 3 ranking classes. In this paper, we're interested in the FNIR vs. R across the different cohorts including the demographic groups, mask/normal facial images, and visual/thermal image domains. The comparison of different curves for each cohorts captures the performance differential at different ranks.

Error tradeoff curves are an approach to visualize the error rates FNIR vs. FPIR at selected intervals of threshold T. At each threshold value T, a corresponding FNIR and FPIR is estimated. Where threshold T determines the amount of acceptance and rejection. FNIR has an inverse relationship with FPIR where as the threshold increases, FNIR decreases and FPIR increases. Error tradeoff curves illustrate the overall performance of the system, specifically the faster and steeper the drop, the better performing system. By plotting each error tradeoff curves against each other for each cohorts, we can visualize the performance differential across both the type I (FPIR) and type II (FNIR) errors.

4.4   Convolutional Neural Network

In this paper, we choose to use two simple deep convolutional neural networks to evaluate the overall facial recognition performance and fairness metric. The purpose of the experiment is not to maximize facial recognition performance but to observe the change in fairness across different modalities: real vs. synthetic images, thermal vs. visual, and demographic groups (age, gender, and ethnicity). In addition, simple CNNs were chosen as an example classifier to illustrate the process of measuring fairness/bias induced by the training process using imbalanced data. Note that the 3-block CNN represents the edge scenario for the dataset used in this paper as the recognition performance is nigh perfect. More sophisticated CNNs, such as InceptionV3 or ArcFace, can be used to marginally increase performance but will have little impact on seeing the changes in fairness.

Table 2 shows the CNN architectures used in this paper. Both CNNs were using the Adam optimizer with default parameters: learning rate $\alpha = 0.001$, $\beta_1 = 0.9$, and $\beta_2 = 0.999$. In order to reduce bias as a result of pre-training, neither CNNs were pre-trained, therefore all the initial weights are randomized. For each subject, 10% of the images were used for training and the remaining 90% were used for testing. For example, given a particular subject with 100 images, 10 images are used for training and 90 images are used for testing. The networks were trained with a batch size of 32 for a total of 10 epochs. This extreme partition of training/testing sets accompanied by low epoch count was chosen to evaluate the role of imbalanced data on facial recognition, specifically assessing fairness in the dataset.

Table 2. 2-Block and 3-Block CNN Architecture.

2-block CNN	3-block CNN
Input $256 \times 256 \times 3$	Input $256 \times 256 \times 3$
64 Conv2D 3×3	64 Conv2D 3×3
64 Conv2D 3×3	64 Conv2D 3×3
Max Pooling	Max Pooling
Batch Normalization	Batch Normalization
128 Conv2D 3×3	128 Conv2D 3×3
128 Conv2D 3×3	128 Conv2D 3×3
Max Pooling	Max Pooling
Batch Normalization	Batch Normalization
	256 Conv2D 3×3
	256 Conv2D 3×3
	Max Pooling
	Batch Normalization
Global Average Pooling	
Fully-Connected	
Softmax Classification	

5   Experimental Results

The experimental study involves the use of two simple 2-block and 3-block convolutional neural networks to perform facial recognition. The performance (TPR, FPR, PPV, NPV), fairness, cumulative match characteristic curves, and error tradeoff curves are assessed using the test set which consists of 90% (38222 images) of the images.

5.1   Performance Across Groups

In this paper, we evaluate the performance and fairness of three different demographic groups: gender, ethnicity, and age. For gender, we separate the dataset into binary male/female classes based on the gender labels provided in the dataset. For ethnicity, we divide it into three categories: Asians (A), Blacks (B), and Caucasians (C), based on the ethnicity labels provided for each subject. Lastly, for age, we split the dataset into 4 age groups (with ages varying from 20 to 57); we provide the distribution of age occurring in the dataset: <25, 25 to 30, 31 to 35, and > 35, based on the reported age of each subject. The occurrence rate for each demographic group are as follows (Table 3):

Table 3. Distribution (%) for different demographic groups: Gender, Ethnicity, and Age.

Gender		Ethnicity			Age			
Male	Female	A	B	C	<25	25-30	31-35	> 35
56.25	43.75	75.00	3.75	21.25	38.75	32.50	12.50	16.25

The distribution for each demographic group shows that the dataset is not balanced, that is, the ratio of male-to-female or A-to-B-to-C is not equal. When an imbalanced dataset is used for training a CNN, it can lead to a biased network. An example of the performance of a biased network is shown in Tables 4 and 5. The 2-block CNN is used as the recognition model with SpeakingFace (unmask, normal) and thermal-mask (mask) datasets used for evaluation. The rows represent the performance based on the different demographic groups such as gender, age, and ethnicity. For this experiment, we show the performance in terms of TPR, FPR, PPV, and NPV measured for visual or thermal images as well as real (unmask, normal) or synthetic (masked) images.

For example, the 2nd row in Table 5 shows an example of a fixed age, 32. This age shows high performance for thermal images regardless of the real or synthetic nature. An interesting disparity is shown when comparing the real and synthetic performance of the visual images using TPR. This disparity for the 32-age demographic group shows a masked TPR rate for a visual image of 30.45%, while the unmasked counterpart reports a TPR of 99.79%. This shows that a simple 2-Block CNN is more capable of distinguishing features from an unmasked subject in comparison to a masked subject.

Since a difference in performance for each class in each demographic group represents a deviation from fairness, we observe that in our experiments, gender seems to deviate less, while age causes a greater deviation; Therefore, age is an attribute that is the most attractive to monitor while assessing fairness in facial recognition algorithms.

5.2   Fairness Across Groups

In this paper, fairness is evaluated across different modalities, including: demographic groups (age, gender, and ethnicity), spectral domains (thermal vs. visual), and synthetic image modification (masked vs. unmasked). We're interested in showing how the framework of training and process of addressing bias can be applied to any recognition system, particularly on synthetic images where masks are synthetically placed on faces.

To address the problem of fairness, we propose 4 pipelines of data inputs: visual-mask, visual-normal, thermal-mask, and thermal-normal images. The controlled variable for each pipeline is the same demographic data according to Table 3 and the same CNN architecture. Each pipeline is evaluated for fairness according to the DPD (Eq. 9) and EOD (Eq. 13) metrics.

Table 6 shows the fairness metrics on the two CNNs. Given the same hyperparameters used for training both CNNs, the 3-Block CNN greatly outperforms the

Table 4. 2-Block CNN Facial Recognition Visual Performance (%) in terms of TPR, FPR, PPV, and NPV.

| | Visual | | | | | | | |
| | Mask | | | | Normal | | | |
	TPR	FPR	PPV	NPV	TPR	FPR	PPV	NPV
Baseline	53.93	0.58	70.94	99.42	83.01	0.22	88.10	99.79
Age:32	30.45	0.00	100.00	99.11	99.79	0.03	97.98	100.00
Age:<25	55.06	0.66	69.55	99.44	79.95	0.22	86.85	99.75
Age:25–30	52.31	0.74	70.31	99.39	83.77	0.33	84.58	99.79
Age:31–35	46.00	0.13	81.29	99.32	82.76	0.09	92.95	99.78
Age:>35	60.57	0.44	67.58	99.51	88.97	0.08	94.38	99.87
Gender:Male	55.70	0.66	70.31	99.44	82.97	0.20	87.38	99.79
Gender:Female	51.66	0.49	71.76	99.39	83.07	0.23	89.03	99.78
Ethnicity:Asian	50.66	0.67	70.97	99.38	81.96	0.26	86.10	99.77
Ethnicity:Caucasian	63.18	0.30	69.96	99.54	86.42	0.07	95.33	99.84
Ethnicity:Black	66.87	0.48	75.95	99.58	84.77	0.23	86.97	99.80

Table 5. 2-Block CNN Facial Recognition Thermal Performance (%) in terms of TPR, FPR, PPV, and NPV.

| | Thermal | | | | | | | |
| | Mask | | | | Normal | | | |
	TPR	FPR	PPV	NPV	TPR	FPR	PPV	NPV
Baseline	69.96	0.38	82.20	99.62	85.82	0.18	89.57	99.82
Age:32	100.00	0.23	85.11	100.00	100.00	0.08	94.19	100.00
Age:<25	62.14	0.22	85.94	99.53	81.92	0.19	88.97	99.77
Age:25–30	73.68	0.47	80.69	99.66	85.63	0.12	92.83	99.82
Age:31–35	73.39	0.48	80.68	99.66	95.78	0.36	81.84	99.95
Age:>35	78.56	0.52	77.45	99.74	87.88	0.14	90.40	99.86
Gender:Male	68.77	0.43	81.05	99.61	88.01	0.13	91.26	99.85
Gender:Female	71.50	0.32	83.67	99.64	83.02	0.23	87.39	99.79
Ethnicity:Asian	66.53	0.36	81.88	99.58	84.34	0.19	89.70	99.80
Ethnicity:Caucasian	78.59	0.50	80.99	99.74	89.08	0.18	87.50	99.87
Ethnicity:Black	89.71	0.06	95.39	99.87	97.05	0.02	98.70	99.96

2-Block CNN. The 3-Block CNN achieves near-perfect recognition performance on the thermal images with a slight decrease in TPR for the visual images. The rows in the table represent the pipeline used for experiments and the column represents the different fairness metrics used.

Demographic Groups: The key observation with respect to the age, gender, and ethnicity demographic groups is its embedded relation to recognition performance. As the TPR approaches 100%, the DPD approaches 0.10%, 0.01%, and 0.01%, for age, gender, and ethnicity, respectively. The approached value should ideally be 0 to indicate that every group's positive rate is equivalent and therefore system contains no false positives. As we decrease the performance of recognition, either by reducing the model learning capacity or increasing noise in an image, we can see a decrease in DPD and EOD. We observe the system is signficantly more biased with respect to age. This is most likely due to the many age groups (24) with a limited amount of training subjects (80). More training samples for each age category is necessary for a more reliable assessment of bias with respect to age as current data reveals one or two subject from each category controls the performance for that age group.

Multi-spectral Domains: The main observation of visual vs. thermal images is that the DPDs are heavily related to the performance of the recognition system. The higher the TPR, the lower the resulting DPDs. One interesting note is that the normal-visual is fairer than the normal-thermal variant across all three demographic groups. Normal-visual has a TPR of 83.01% against normal-thermal's TPR of 85.82% where the corresponding visual's DPD of 27.86%, 0.13%, and 4.55% against thermal's DPD of 34.02%, 4.90%, and 12.64%. This illustrates that even though there is a relationship between performance and fairness, higher performance does not directly result in a lower fairness metric.

Synthetic Mask Application: The fairness of the system degrades when using applying synthetic masks. This is observed when looking at the masked/unmasked images of the visual domain. Normal-visual has a DPD of 27.86%, 0.13%, and 4.55% (age, gender, and ethnicity groups) which is increased to 75.73%, 3.72%, and 16.06% when wearing masks. An increase in DPD indicates the system is producing less fair results. Because synthetic modification is additional processing applied on top of already existing images, this modification process may introduce an additional layer of biases. This is likely one reason why the masked variant of the same image is more biased.

Evaluating fairness across different modalities is one critical process to understanding recognition systems. In this section, we show how demographic groups, multi-spectral domains, and synthetic images can be analyzed to show the level of fairness. Synthetic images are particularly of interest as it produces additional biases if the method of application is unfair. The emerging use of synthetic images to train recognition systems is a growing concern as the system can be seeded with additional bias that is hidden. The process of using fairness metrics is one method to reveal these hidden dangers.

Table 6. Facial Recognition Fairness Evaluation (%).

	DPD			EOD		
	Age	Gender	Ethnicity	Age	Gender	Ethnicity
2-Block CNN						
Masked-Visual	75.73	3.72	16.06	30.32	11.27	18.52
Normal-Visual	27.86	0.13	4.55	61.93	3.64	45.07
Masked-Thermal	49.06	3.19	23.10	31.48	11.43	29.31
Normal-Thermal	34.02	4.90	12.64	49.90	3.45	30.19
3-Block CNN						
Masked-Visual	3.72	0.21	0.51	100.00	5.50	19.06
Normal-Visual	2.47	0.05	0.82	100.00	5.65	14.25
Masked-Thermal	0.10	0.01	0.01	100.00	2.79	1.65
Normal-Thermal	0.10	0.01	0.01	100.00	0.01	1.65

5.3 Fairness via Performance Differentials

In this paper, we have addressed the problem of trust in the autonomous system via performance and fairness. By evaluating the performance differentials, we're able to directly see the performance behaviors across the different demographic groups, different spectral domains, and the influence of applying synthetic masks.

Table 7. Facial Recognition Performance (Rank-1 Rates, %).

	2-Block CNN				3-Block CNN			
	TPR	FPR	PPV	NPV	TPR	FPR	PPV	NPV
Masked-Visual	53.93	0.58	70.94	99.42	99.54	0.01	99.58	99.99
Normal-Visual	83.01	0.22	88.10	99.79	99.87	0.00	99.89	100.00
Masked-Thermal	69.96	0.38	82.20	99.62	99.99	0.00	99.99	100.00
Normal-Visual	85.82	0.18	89.57	99.82	99.99	0.00	99.99	100.00

Table 7 shows the summarized rank-1 performance for the 2-block and 3-block CNNs. Each column in the table represents a metric used to capture the CNNs' average identification performance. Hit rate (TPR) and miss rate (FPR) are 2 common metrics used to measure performance. In this experiment, we show that the performance of masked images (53.93%) on average is significantly worse than their unmasked (83.01%) counterparts in terms of both higher hit rate and lower miss rate. In addition, the performance of the thermal images is also higher than the visual images. Since the CNNs, thermal-to-visual image ratio, and mask-to-unmasked image ratio are controlled, these two observations illustrate that the CNNs are inherently biased towards a specific type of image, specifically the CNNs learn significantly better features for the identification of thermal images. The CNNs are also likely to be biased for unmasked images due to the higher performance but due to the nature of synthetic image modification, the bias may be the result of the synthetic modification instead of the identification process.

Figure 3 shows the CMC curves for mask and unmasked facial images across different demographic groups. For each figure, there are 4 curves representing the masked/unmasked images in both the visual and thermal domains. Performance is represented in the form of FNIR as a function of rank, where the lower FNIR the better. The key observation is the poor performance of the visual-masked curve amongst all demographic groups. The visual-mask curve obtains a significantly higher FNIR in almost all aspects with the exception of Ethnicity-B where masked-thermal is slightly worse after rank-7. The CMC curves provide evidence that the recognition system is unfair towards specific demographic groups, in particular, Ethnicity-B has significantly lower FNIR compared to other ethnicity groups.

Figure 4 shows the identification error rates for different demographic groups. For each figure, the error rates for masked/unmasked images across visual and thermal domains are shown. The primary observation is the Ethnicity-B group where the FNIR vs. FPIR is significantly lower with respect to other ethnicity groups. The Ethnicity-A group is relatively similar to the Baseline, most likely due to the dataset consisting of 75% Asian ethnicity. The key outlier is the visual-mask error curve which is consistently worse than any other combinations. This outlier supports the evidence of bias in the recognition system, specifically, the CNNs are unable to extract good identifiable features for the masked variant of visual images in contrast to other variants.

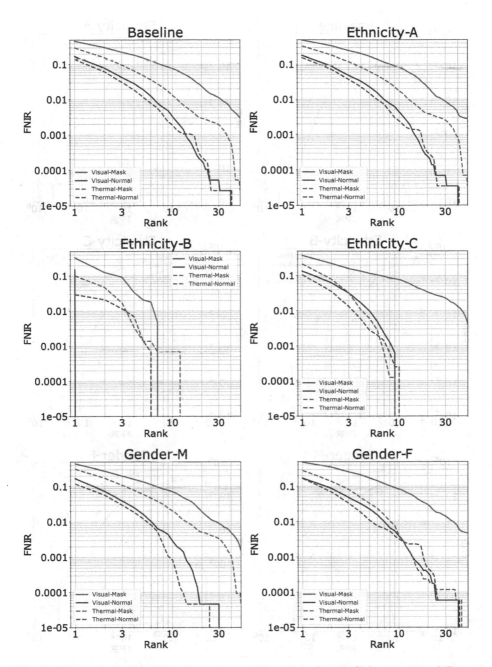

Fig. 3. False negative identification rates as a function of ranks (false negative differentials) between the mask and normal face images for visual and thermal domains. Solid lines represent the visual domain, dashed lines represent the thermal domain.

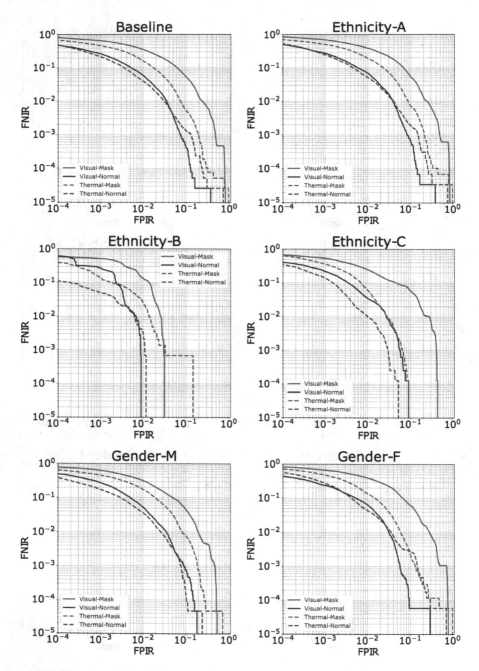

Fig. 4. False negative identification rates as a function of false positive identification rate between the mask and normal face images for visual and thermal domains. Solid lines represent the visual domain, dashed lines represent the thermal domain.

6   Summary and Conclusions

In order for a user to establish trust in an autonomous system, the system must provide reliable and fair results. Our proposed framework demonstrates how these pillars of trust must be satisfied in order to build trustworthiness in an autonomous system:

1. **Reliability**: The recognition system must be accurate in terms of high hit rate and low miss rate for it to be reliable. An unreliable system is not trustworthy as the results of the system cannot be used for any decision-making. We measure the reliability of the system to have an average rank-1 hit rate (true positive rate, TPR) of 83.01% and 99.87% using 2-block and 3-block convolutional neural networks (CNNs), respectively.
2. **Fairness**: The recognition system must be fair so that it is unbiased towards any group. A biased system is untrustworthy as specific groups can be targeted. We examined the fairness of the proposed recognition system across different modalities.

Bias is one common denominator for unfair systems. In this paper, we report several modalities that may contain bias, including demographic groups, spectral domains, and real vs. synthetic image modification.

Demographic Groups: The collection process of a dataset may lead to an imbalanced dataset which may result in demographic biases, specifically biased cohorts in the form of demographic groups such as gender, ethnicity, and age.

- These biases can lead to an unfair decision as they can deeply seed the training process in a machine learning algorithm with bias.
- In our experiments, we show that as the performance improves, the DPDs increase proportionally to the number of classes in the demographic group. We report a higher deviation in age compared to gender.
- The level of fairness (DPD) across age, gender, and ethnicity is 0.10%, 0.01%, and 0.01% for a 3-block CNN with a TPR of 99.99%.
- Performance differentials show that demographic groups such as different ethnicities contain a significant bias. For example, the TPR for Asians is 84.34% which is significantly lower than for Blacks with a TPR of 97.05%.

Multi-spectral Domains: The use of pre-trained deep-neural networks results in a bias towards visual (RGB) images due to the predominant training on the ImageNet images. Other spectral images such as thermal (infrared) obtain relatively good results but are most often not as accurate as their visual counterparts.

- We chose to use randomized weights for the CNNs as the starting point to remove the impact of pre-training bias.
- In our experiments, thermal images are better for facial identification as it provides higher hit rates and lower miss rates.

– Thermal images obtain a TPR of 85.52% for unmasked images which is greater than the visual image's TPR of 83.01%. Taking into account age, gender, and ethnicity, thermal images obtain a DPD of 34.02%, 4.9%, and 12.64% which are comparably higher than the visual images DPD of 27.86%, 0.13%, and 4.55%, indicating that lower the TPR of visual images produces fairer results.
– Performance differentials indicate a preference for thermal images. For masked images, thermal images retrieve a TPR of 69.96% as opposed to visual images TPR of 53.93%.

Synthetic Image Modifications: The application of synthetic modifications to an image may lead to bias.

– Bias comes in the form of recognition performance, particularly, through the use of synthetic images, the recognition performance is much better or worse than its unmodified version.
– In our experiments, we show that the synthetic mask version of the same facial images has a higher miss rate and a lower hit rate with a bigger decrease in performance for visual images in comparison to thermal images.
– The TPR for a normal unmasked visual image is 83.01% but is decreased to 53.93% when a synthetic mask is applied. Similarly, the TPR for a normal unmasked thermal image is 85.85% but is decreased to 69.96% when a mask is applied. As a result of the change in performance, the DPD across most demographic groups also decreases.
– Performance differentials support the founding that by applying synthetic masks, the overall performance of the system decreases.

Further study should involve building a combined real and synthetic dataset, in which the synthetic images would be used to augment the classes with few samples to create an overall more balanced dataset. The related measures such as risk and trust [16, 22], shall be also evaluated, aiding the assessment of fairness.

Acknowledgments. This work was supported in part by the Department of National Defence's Innovation for Defence Excellence and Security (IDEaS) program, Canada; in part by the Social Sciences and Humanities Research Council of Canada (SSHRC) through the Grant "Emergency Management Cycle-Centric R&D: From National Prototyping to Global Implementation" under Grant NFRF-2021-00277; in part by the University of Calgary under the Eyes High Postdoctoral Match-Funding Program.

References

1. Grother, P., Ngan, M., Hanaoka, K.: Face recognition vendor test (FRVT) part 2: identification. In: National Institute of Standards and Technology (NIST), Gaithersburg, MD, USA, Tech. Rep. NISTIR 8271 (2019)
2. Valdivia, A., Corbera-Serrajòrdia, J., Swianiewicz, A.: There is an elephant in the room: Towards a critique on the use of fairness in biometrics. arXiv preprint arXiv:2112.11193. (2021)

3. Cohen, L., Lipton, Z.C., Mansour, Y.: Efficient candidate screening under multiple tests and implications for fairness. arXiv preprint arXiv:1905.11361 (2019)
4. Dwork, C., Hardt, M., Pitassi, T., Reingold, O., Zemel, R.: Fairness through awareness. In: The 3rd Innovations in Theoretical Computer Science Conference, pp. 214–226 (2012)
5. Howard, J.J., Rabbitt, L.R., Sirotin, Y.B.: Human-algorithm teaming in face recognition: How algorithm outcomes cognitively bias human decision-making. Plos One 15(8), e0237855 (2020)
6. Hugenberg, K., Wilson, J.P., See, P.E., Young, S.G.: Towards a synthetic model of own group biases in face memory. Visual Cogn. 21(9–10), 1392–1417 (2013)
7. Verma, S., Rubin, J.: Fairness definitions explained. In: IEEE/ACM International Workshop on Software Fairness, pp. 1–7 (2018)
8. Ntoutsi, E., et al.: Bias in data-driven artificial intelligence systems-an introductory survey. Wiley Interdisc. Rev. Data Min. Knowl. Discov. 10(3), e1356 (2020)
9. Mehrabi, N., Morstatter, F., Saxena, N., Lerman, K., Galstyan, A.: A survey on bias and fairness in machine learning. ACM Comput. Surv. 54(6), 1–35 (2021)
10. de Freitas Pereira, T., Marcel, S.: Fairness in biometrics: a figure of merit to assess biometric verification systems. IEEE Trans. Biometrics Behav. Identity Sci. 4(1), 19–29 (2021)
11. Rathgeb, C., Drozdowski, P., Damer, N., Frings, D.C., Busch, C.: Demographic fairness in biometric systems: What do the experts say? arXiv preprint arXiv:2105.14844 (2021)
12. Krishnan, A., Almadan, A., Rattani, A.: Probing fairness of mobile ocular biometrics methods across gender on VISOB 2.0 dataset. In: International Conference on Pattern Recognition, pp. 229–243 (2021)
13. Merler, M., Ratha, N., Feris, R.S., Smith, J.R.: Diversity in faces. arXiv preprint arXiv:1901.10436 (2019)
14. Drozdowski, P., Rathgeb, C., Busch, C.: Demographic fairness in face identification: The watchlist imbalance effect. arXiv preprint arXiv:2106.08049 (2021)
15. Terhörst, P., Kolf, J.N., Damer, N., Kirchbuchner, F., Kuijper, A.: Post-comparison mitigation of demographic bias in face recognition using fair score normalization. Pattern Recogn. Lett. 140, 332–338 (2020)
16. Yanushkevich, S., Stoica, A., Shmerko, P., Howells, W., Crockett, K., Guest, R.: Cognitive identity management: synthetic data, risk and trust. In: International Joint Conference on Neural Networks, pp. 1–8 (2020)
17. Feng, L., Wiltsche, C., Humphrey, L., Topcu, U.: Synthesis of human-in-the-loop control protocols for autonomous systems. IEEE Trans. Autom. Sci. Eng. 13(2), 450–462 (2016)
18. Liang, Y., Samtani, S., Guo, B., Yu, Z.: Behavioral biometrics for continuous authentication in the internet-of-things era: an artificial intelligence perspective. IEEE Internet Things J. 7(9), 9128–9143 (2020)
19. Peternel, L., Tsagarakis, N., Caldwell, D., Ajoudani, A.: Robot adaptation to human physical fatigue in human-robot co-manipulation. Auton. Robots 42(5), 1011–1021 (2018)
20. Pollak, A., Paliga, M., Pulopulos, M.M., Kozusznik, B., Kozusznik, M.W.: Stress in manual and autonomous modes of collaboration with a cobot. Comput. Hum. Behav. 112, 106469 (2020)
21. Martínez, A., Belmonte, L.M., García, A.S., Fernández-Caballero, A., Morales, R.: Facial emotion recognition from an unmanned flying social robot for home care of dependent people. Electronics 10(7), 868 (2021)

22. Yanushkevich, S., Reitinger, N., Stoica, A., Oliveira, H.C.R., Shmerko, V.: Inverse biometrics: privacy, risks, and trust. In: Jajodia, S., Samarati, P., Yung, M. (eds.), Encyclopedia of Cryptography, Security and Privacy. Springer, Berlin, Heidelberg (2021). https://doi.org/10.1007/978-3-642-27739-9_1505-1
23. Lai, K., Oliveira, H.C., Hou, M., Yanushkevich, S.N., Shmerko, V.P.: Risk, trust, and bias: causal regulators of biometric-enabled decision support. IEEE Access **8**, 148779–148792 (2020)
24. Granger, E., Gorodnichy, D.: Evaluation methodology for face recognition technology in video surveillance applications. Defence R & D Canada (2014)
25. Phillips, P.J., Moon, H., Rizvi, S.A., Rauss, P.J.: The FERET evaluation methodology for face-recognition algorithms. IEEE Trans. Pattern Anal. Mach. Intell. **22**(10), 1090–1104 (2000)
26. Gunning, D., Stefik, M., Choi, J., Miller, T., Stumpf, S., Yang, G.-Z.: XAI-Explainable artificial intelligence. Sci. Robot. **4**(37), eaay7120 (2019)
27. Rudin, C.: Stop explaining black box machine learning models for high stakes decisions and use interpretable models instead. Nat. Mach. Intell. **1**(5), 206–215 (2019)
28. Arrieta, A.B., et al.: Explainable artificial intelligence (XAI): concepts, taxonomies, opportunities and challenges toward responsible AI. Inf. Fusion **58**, 82–115 (2020)
29. Barocas, S., Hardt, M., Narayanan, A.: Fairness in machine learning. NIPS Tutorial **1**, 2 (2017)
30. Grother, P., Ngan, M., Hanaoka, K.: Face recognition vendor test (FRVT) part 3: demographic effect. In: National Institute of Standards and Technology (NIST), Gaithersburg, MD, USA, Tech. Rep. NISTIR 8280 (2019)
31. Hardt, M., Price, E., Srebro, N.: Equality of opportunity in supervised learning. Adv. Neural Inf. Process. Syst. **29**, 3323–3331 (2016)
32. Kusner, M.J., Loftus, J., Russell, C., Silva, R.: Counterfactual fairness. In: Advances in Neural Information Processing Systems, vol. 30 (2017)
33. Qian, S., et al.: Are my deep learning systems fair? An empirical study of fixed-seed training. Adv. Neural Inf. Process. Syst. **34**, 30211–30227 (2021)
34. Liu, H., et al.: Trustworthy AI: a computational perspective. ACM Trans. Intell. Syst. Technol. **14**(1), 1–59 (2022)
35. Jeevan, G., Zacharias, G.C., Nair, M.S., Rajan, J.: An empirical study of the impact of masks on face recognition. Pattern Recogn. **122**, 108308 (2022)
36. Lai, K., Queiroz, L., Shmerko, V., Sundberg, K., Yanushkevich, S.N.: Post-pandemic follow-up audit of security checkpoints. IEEE Access, 1–18 (2023)
37. Talahua, J.S., Buele, J., Calvopiña, P., Varela-Aldás, J.: Facial recognition system for people with and without face mask in times of the covid-19 pandemic. Sustainability **13**(12), 6900 (2021)
38. Wang, M., Deng, W., Hu, J., Tao, X., Huang, Y.: Racial faces in the wild: reducing racial bias by information maximization adaptation network. In: IEEE/CVF International Conference on Computer Vision, pp. 692–702 (2019)
39. Yu, J., Hao, X., Cui, Z., He, P., Liu, T.: Boosting fairness for masked face recognition. In: IEEE/CVF International Conference on Computer Vision, pp. 1531–1540 (2021)
40. Wang, Z., et al.: Towards fairness in visual recognition: effective strategies for bias mitigation. In: IEEE/CVF Conference on Computer Vision and Pattern Recognition, pp. 8919–8928 (2020)
41. Abdrakhmanova, M., Kuzdeuov, A., Jarju, S., Khassanov, Y., Lewis, M., Varol, H.A.: SpeakingFaces: a large-scale multimodal dataset of voice commands with visual and thermal video streams. Sensors **21**(10), 3465 (2021)

42. Queiroz, L., Oliveira, H., Yanushkevich, S.: Thermal-mask-a dataset for facial mask detection and breathing rate measurement. In: International Conference on Information and Digital Technologies (IDT), pp. 142–151 (2021)
43. Deng, J., Guo, J., Ververas, E., Kotsia, I., Zafeiriou, S.: RetinaFace: single-shot multi-level face localisation in the wild. In: IEEE/CVF Conference on Computer Vision and Pattern Recognition, pp. 5203–5212 (2020)
44. Sun, K., et al.: High-resolution representations for labeling pixels and regions. arXiv preprint arXiv:1904.04514 (2019)
45. Bolle, R.M., Connell, J.H., Pankanti, S., Ratha, N.K., Senior, A.W.: The relation between the ROC curve and the CMC. In: IEEE Workshop on Automatic Identification Advanced Technologies, pp. 15–20 (2005)

An Autonomous Fake News Recognition System by Semantic Learning and Cognitive Computing

Yingxu Wang[✉] and James Y. Xu

Lab for Computational Intelligence and Cognitive Systems, International Institute of Cognitive Informatics and Cognitive Computing, Department of Electrical and Software Engineering, Schulich School of Engineering (SSE) and Hotchkiss Brain Institute (HBI), University of Calgary, 2500 University Drive NW, Calgary, AB T2N 1N4, Canada
{yingxu,yifan.xu1}@ucalgary.ca

Abstract. It has been well understood that fake news recognition is a persistent challenge to cognitive computing and autonomous systems in general, and to *Artificial Intelligence* (AI), machine knowledge learning, and computational linguistics in particular. This work develops an *Autonomous Fake News Recognition* (AFNR) system by cognitive computing theories underpinned by *Intelligent Mathematics* (IM) such as concept algebra and semantic algebra. A training-free methodology and a formal algorithm for *Differential Sematic Analysis* (DSA) are designed in *Real-Time Process Algebra* (RTPA) and implemented in MATLAB. The AFNR system is implemented in the Anaconda environment with Python, the natural language toolkit (NLTK), and an English parser – Spacy. Compared to the classical data-driven neural network methodologies, AFNR and DSA have demonstrated a significant improvement against the level of accuracy over the randomly selected and large-scale benchmark of a fake news database. The DSA methodology for fake news recognition has enabled autonomous machine knowledge learning and semantic comprehension towards differential and robust semantic analyses for fake news in natural languages. The AFNR system has reached an accuracy level of 70.1%, which over performs the top ranked teams in DataCup'19 with the highest reported accuracy of 55.0%.

Keywords: Fake news · Recognition · Semantic analysis · Knowledge learning · Intelligent mathematics · Cognitive systems · Cognitive computing · Computational linguistics · Autonomous systems

1 Introduction

Fake news is deliberately untrue or partially misleading news that is inconsistent to real events and background facts [1–16]. Autonomous fake news detection and recognition are not only a fundamental demand in the electronic information era, but also a persistent challenge to cognitive computing theories [1, 15, 33] and AI technologies [16–20]. Along with the explosively increasing volume of electronic social media, fake news has been recognized as one of the major risks that undermines media integrity and their cognitive trustworthiness [2, 10, 12, 21, 23, 48].

M. Gavrilova et al. (Eds.): *Transactions on Computational Science XL*, LNCS 13850, pp. 88–109, 2023.
https://doi.org/10.1007/978-3-662-67868-8_6

Although a variety of technologies and conceptual solutions have been proposed in the literature, the majority of them are focusing on the analysis and extraction of textural features in articles where neural network powered machine learning technologies are utilized to recognize the trustworthiness of news contents [5–9, 21, 22]. Because of the lack of rigorous theories, semantic comprehension methodologies, and cognitive computing tools, the study on theoretical foundations for fake news recognition has left behind the increasing demands to trustworthy cognitive computing [23].

According to *concept algebra* [17] and *sematic algebra* [18] in *Intelligent Mathematics* (IM) [1], the semantics of natural languages expressions embodies the denotational intention and extension of language expressions at the word, phrase, sentence, paragraph, and essay levels, which are carried by syntactical entities (nouns or noun phrases), behaviors (verbs or verb phrases), and modifiers (adjectives, adverbs, etc.) [18]. The cognitive linguistic framework of cognitive computing [15, 20, 33–47] forms a theoretical foundation for fake news recognition by machine knowledge learning [24] and differential semantic analyses [18].

Definition 1. The *universe of discourse U* of human behaviors in language expressions is a 5-dimentional semantic space, where J denotes the subject of a behavior, O the object, A the action, event, or behavior, S the space (location), and T the time point when the event or action occurs in the semantic space.

Based on Definition 1, the semantic space of human behavioral expression in general and news reporting in particular are illustrated in Fig. 1.

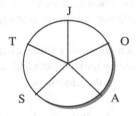

Fig. 1. The semantic space of behavioral processes for news modeling.

This work presents a cognitive computing system for *Autonomous Fake News Recognition* (AFNR) by semantic learning methodologies in cognitive computing [1, 15, 18, 33]. A set of mathematical and semantic models of fake news is created in Sect. 2 encompassing the truth models of declarative semantics and behavioral semantics, as well as the differential semantic model of language expressions. Based on the cognitive computing models, the design of the AFNR system is formally described in Sect. 3 supported by a claim and fact parsing algorithm, a differential semantic analysis algorithm, and a semantic fake news recognition algorithm. Section 4 demonstrate the implementation of the AFNR tool that enables a large set of experiments by cognitive computing [33]. The AFNR system has reached a significant accuracy level of 70.1%, which over performs the top ranked teams in DataCup'19 with a reported accuracy record of 55% [23] that indicates a marginal result a little bit better than those of random guesses.

2 Semantic Models for Differential Semantic Analyses (DSA)

For fake news recognition, a reported claim is considered as false if its semantics is inconsistent to its associated background facts and their syntactical constrains. Such DSA methodology requires a rigorous formal model to enable the AFNR system. According to *concept algebra* [17] and *semantic algebra* [18], the semantic of a natural language sentence may consist of three types of semantics known as the entity, behavior, and modifier.

Semantics is the composed meaning of language expressions and perceptions at the levels of words, phrase, sentence, paragraph, and essay from the bottom up. The basic unit of formal semantics is a target concept or a combined phrase in natural language expressions [18]. According to semantic algebra, the semantic model of fake news is designed in two categories that manipulate the semantics of the *to-be* (noun phrase) and *to-do* (verb phrase) structures [17, 18].

2.1 The Formal Model of Declarative Semantics

A formal definition of the to-be semantic, which is expressed in the form of formal concept, is shown as follows.

Definition 2. The formal *semantics of a conceptual entity E*, $\Theta(E)$, is represented by a formal concept C_E denoted by $\models$:

$$\Theta(E) \triangleq \Theta(E = C_E) \models C_E(A, O, R^c, R^i, R^o) \tag{1}$$

where C_E is specified by the *intension* (attributes A) and *extension* (objects O) of the concept E as well as its internal/input/output relations (R^c, R^i, R^o), respectively.

According to Eq. (1), the semantics of entities elicited from any natural language sentence can be formally denoted by their objects and attributes where the former often refers to the nouns and the latter are the modifiers.

2.2 The Formal Model of Behavioral Semantics

In the previous section, the *to-be* semantics in semantic algebra infers the meaning of an equivalent relation between an unknown and a known entity represented by a pair of concepts. However, the semantics of a behavior is embedded by a *to-do* semantics that denoting the meaning of the action of a person/entity as a behavioral process, which may be formally modeled by *Real-Time Process Algebra* (RTPA) [25].

Definition 3. The formal *semantics of a behavior B*, $\Theta(B)$, in U is a mapping from the subject J to the object O as represented by the formal process P_B to embody the verb V, denoted by $|>$:

$$\Theta(B) \triangleq \Theta(J \xrightarrow{P_B} O) \mathrel{|>} P_B (J, O, A, S, T) \tag{2}$$

where P_B denotes the action process A of the verb as well as its subject (J), object (O), space (S), and time (T).

The semantic space of behavioral processes has been illustrated in Fig. 1 as a 5-D hyperstructure. Based on Definition 3, the semantics of the second category of news expression for behavioral description is derived as follows.

Definition 4. The *semantic composition* of a to-do sentence S_d, $\Theta(S_d)$, is a transformation of a composite behavior semantics of a verb phrase $\Theta(F_B)$ from the subject J to the object O, denoted by $\Theta(B: J \mid> O)$:

$$\Theta(S_d) \triangleq \Theta(B : J \mid> O)$$

$$= \mathop{\breve{U}}_{\substack{B \\ i=1}}^{n} (\Theta(F_{B_i}) \mid \Theta(J) \to \Theta(O)) \tag{3}$$

$$= \mathop{R}_{i=1}^{n} (P_{B_i} \mid J \to O)$$

$$= \mathop{R}_{i=1}^{n} P_{B_i}(J, O, V_i, S_i, T_i)$$

where O may be default or implied in special cases.

Based on Eqs. 2 and 3, the to-do semantic of a sentence can be rigorously denoted as a tuple that invokes two entity phrases (subject and object), a verb phrase, and a complementary phrase for space and time, where the entity phrases are represented by formal concepts as described in Eq. (1).

2.3 The Differential Semantic Model (DSM) of Language Expressions

On the basis of Definitions 2 through 4, the semantic model of DSA is described in Fig. 2 as a rigorous RTPA structure model DSA|SM, where the symbol |Ξ denotes a type of an *array list* which may contain an empty set or a set of finite elements. The associated functional model of DAS, DSA|PM, will be introduced in Sect. 3.

DSA|SM models the structure of a news claim and associated facts stored in File|SM where the Claim|S may be linked to multiple Fact|S in type string (|S). The tokens of words in a claim or a factual sentence are classified into 13 *part-of-speech* (POS) categories denoted as JP|Ξ(subject), VP|Ξ(verb), OP|Ξ(object), MX|Ξ(modifiers), Stc|Ξ(sentence), as well as PC|Ξ, TC|Ξ, XC|Ξ for different complementary clauses as defined in Fig. 2. The notation MX|Ξ refers to a pair of front and post modifiers for subject, verb and object such as MJ1|Ξ (front subject modifier) and MJ2|Ξ(post subject modifier). However, POS|SM is the key structure to be used throughout the algorithm.

PC|Ξ and TC|Ξ are sets of tokens for space or time variables. They usually appear in a sentence in the form of clauses or prepositional phrases. Detailed implementations for extracting TC|Ξ and PC|Ξ will be described in StcComposition|PM in Sect. 3.1. Stc|SM represents the structure model of an arbitrary sentence in natural language. According to the *Deductive Grammar of English* (DGE) [16], Stc|SM can be deductively resolved and analyzed at the level of JP, VP and OP with their front and post modifiers, as well as a complementary clause CC|SM. All elicited POS components are retained in Stc|SM for differential semantic analysis in DSA|PM in Sect. 3.2.

$$\mathbf{DSA}|SM \triangleq (<Claim|SM \triangleq \overset{n|N}{\underset{i=1}{R}} Claim(i)|S >,$$

$$<Fact|SM \triangleq \overset{F_i|N}{\underset{k=1}{R}} Fact(i,k)|S >,$$

$$<File|SM \triangleq \overset{C|N}{\underset{i=1}{R}} \overset{F_i|N}{\underset{k=1}{R}} File(i)|S.Fact(k)|SM>,$$

$$<POS\_Schema|SM \triangleq \overset{C|N}{\underset{i=1}{R}} \overset{F_i|N}{\underset{k=1}{R}} POS(i,k)|SM >, \text{ // Part of speach}$$

$$<POS|SM \triangleq \overset{C|N}{\underset{i=1}{R}} \overset{F_i|N}{\underset{k=1}{R}} \overset{13}{\underset{j=1}{R}} POS(i,k,j)|\Xi >$$

$$= \overset{C|N}{\underset{i=1}{R}} \overset{F_i|N}{\underset{k=1}{R}} POS(i,k)|SM. \overset{13}{\underset{j=1}{R}} \{(1,JP|\Xi),\ (2,VP|\Xi),\ (3,OP|\Xi),$$

$$(4,PC|\Xi),\ (5,TC|\Xi),\ (6,XC|\Xi),$$

$$(7,Stc|\Xi),\ (8,MJ1|\Xi),\ (9,MJ2|\Xi),$$

$$(10,MV1|\Xi),\ (11,MV2|\Xi),\ (12,MO1|\Xi),$$

$$(13,MO2|\Xi)\}$$

$$<Stc|SM \triangleq \{[MJ1|\Xi \mid JP|\Xi \mid MJ2|\Xi] \mid [MV1|\Xi \mid VP|\Xi \mid MV2|\Xi] \mid$$

$$[MO1|\Xi \mid OP|\Xi \mid MO2|\Xi],\ CC|\Xi\}$$

$$<CC|SM \triangleq TC|\Xi \cup PC|\Xi \cup XC|\Xi \quad \text{// Compliment clause}$$

$$<TC|\Xi \triangleq \{when\ (time),\ at, in, during,\ before, after, from\ ...\ to\ ...\} >$$

$$<PC|\Xi \triangleq \{where\ (space),\ at, in,...\} >$$

$$<XC|SM \triangleq CC|\Xi \setminus (TC|\Xi \cup SC|\Xi)$$

$$<NegW|\Xi \triangleq \{no, not, never, few, hardly, scarcely, barely, narrowly, rarely,$$

$$little, seldom, seldomly\}\ ...\} >,$$

$$<DepTree|SM \triangleq \overset{\#Token|N}{\underset{t=1}{R}} (<Token(t)|S \mid 1 \le Token(t)|S \le 255 >,$$

$$<Index(t)|N \mid Index(t)|N = t|N >,$$

$$<POSTag(t)|S \mid POSTag(t)|S \in UPT|\Xi >$$

$$<Dep(t)|S \mid Dep(t)|S \in SynDep|\Xi >,$$

$$<Assoc(t)|N \mid 1 \le Assoc(t)|N \le \#Tokens|N >$$

$$)>$$

$$< Sim|SM \triangleq \overset{F_i|N}{\underset{k=1}{R}} (Sim(k)|N \mid Sim(k)|N \in \{(0, Negative),\ (1, Positive),$$

$$(2, Partial)\}) >$$

$$)$$

Fig. 2. The Semantic Model of DSA

In order to detect any contradicted semantics, a set of negative keywords NegW|Ξ is pre-defined. A dependency tree structure DepTree|SM is created to store the derived meaning of a natural language sentence by Spacy [26]. The finally generated semantic similarity result will be saved in Sim|SM.

TC|Ξ and PC|Ξ are sets of tokens for time and space variants. They usually appear in a sentence in form of clauses or prepositional phrases. The detailed implementation for extracting TC|Ξ and PC|Ξ is described in StcComposition|PM in Sect. 3.1. Stc|SM represents structure of an arbitrary sentence in natural language. According to DGE [16], Stc|SM can be deductively resolved and analyzed at the level of JP, VP and OP with their front and post modifiers, as well as a complementary clause CC|SM. All elicited POS components will be retained in Stc|SM for differential semantic analysis in DSA|PM in Sect. 3.2.

In order to determine any contradicted semantics, a set of negative keywords NegW|Ξ is pre-defined. A dependency tree structure DepTree|SM is created to store the derived meaning of a natural language sentence by Spacy [26].

3   Design of the AFNR System

The *Algorithm of Differential Semantic Analysis* (DSA) of DSA is formally described in Fig. 4 in three processes. Process (a) conducts two parsing processes Parse|PM where each function is followed by a sentence composition process StcComposition|PM. Parse|PM adopts a claim and its associated facts for eliciting the part-of-speech (POS) components, while Stc|S is generated based on function StcComposition|PM. Process (b) carries a differential semantic analysis process that generates matching scores Match|N for the current claim against each of its associated facts. Then, Process (c) calculates a final similarity value Sim|N based on the accumulated Match|N across all facts.

The DSA algorithm is designed using double-layer iterations to process each queried claim and its associated facts. The inner loop denoted by $\overset{F_i|N}{\underset{k=1}{R}}$ conducts Processes (a) and (b) for every fact stored in File|SM against Claim|S. The outer loop is carried out by Process (c) that determines Sim|N based on Match|N generated in the loop. Then, the loop denoted by $\overset{C|N}{\underset{i=1}{R}}$ (...) analyzes the claim of a suspect news identified by ID|N.

3.1   Claims and Facts Parsing

The process Parse|PM is a built-in function in DSA|PM for both claim and fact parsing as invoked in Process (s) in Fig. 4. Parse|PM is implemented on the platform of Spacy, which serves as a logical controller for switching the data flow between a fetched claim and the associated facts determined by the Boolean variable Claim|L. It returns the POS components as the analysis results to the main DSA|PM process.

The process Parser|PM embedded in Fig. 3 elicits the main POS components (JP, VP, OP) and their modifiers (MP, MV, MO) as defined in Fig. 2 for a given claim or fact sentence as shown in Figs. 5 and 6. Parser|PM invokes a function call from the Spacy library [26] to acquire a dependency tree structure, DepTree|SM, against the input string Sentence|S at the first place. DepTree|SM is a group of token instances derived from each word in the Sentence|S string. Each instance uses attributes Token|S, Index|N, POSTag|S, Dep|S and Assoc|N for holding information of the context, ID, POS tag, dependency, and associated ID of the token.

$$\textbf{Parse}|PM(<\textbf{I}::\; i|N,\; Claim|L,\; F_i|N,\; \overset{F_i|N}{\underset{k=1}{R}}\; Sentence(i,k)|S>;\; <\textbf{O}::\; \overset{n|N}{\underset{i=1}{R}}\;\overset{F_i|N}{\underset{k=1}{R}}\;\overset{13}{\underset{j=1}{R}}\; POS(i,k,j)|\Xi >;$$

$$<\textbf{H}::\; \overset{n|N}{\underset{i=1}{R}}\;\overset{F|N}{\underset{k=1}{R}}\; POS(i,k)|SM,\; \overset{C|N}{\underset{i=1}{R}}\;\overset{F|N}{\underset{k=1}{R}}\; File(i)|S.Fact(k)|S>)$$

$\{\rightarrow (\; \blacklozenge\; Claim|L = T|L \quad // \text{Processing a claim}$

$\qquad \rightarrow N_s|N := 1$

$$\rightarrowtail \textbf{Parser}|PM(<\textbf{I}::\; i|N,\; Claim(i)|S,\; N_s|N>;\; <\textbf{O}::\; \overset{13}{\underset{j=1}{R}}\; POS(i,1,j)|\Xi;$$

$$<\textbf{H}::\; \overset{6}{\underset{j=1}{R}}\; Claim(i,j)|\Xi,\; \overset{N_s|N}{\underset{i=1}{R}}\;\overset{F_i|N}{\underset{k=1}{R}}\; POS(i,k)|SM,$$

$$\overset{C|N}{\underset{i=1}{R}}\;\overset{F|N}{\underset{k=1}{R}}\; File(i)|S.Fact(k)|S>)$$

$\;|\sim$

$\qquad // \text{Processing a set of facts}$

$\qquad \rightarrow N_s|N := F_i|N$

$$\rightarrow \overset{F_i|N}{\underset{k=1}{R}}\; (Fact(i,k)|S := File(i)|S.Fact(k)|S)$$

$$\rightarrow \overset{F|N}{\underset{k=1}{R}}\; (\rightarrowtail \textbf{Parser}|PM(<\textbf{I}::\; Fact(i,k)|S,\; N_s|N>;\; <\textbf{O}::\; \overset{13}{\underset{j=1}{R}}\; POS(i,k,j)|\Xi;$$

$$<\textbf{H}::\; \overset{F|N}{\underset{k=1}{R}}\;\overset{6}{\underset{j=1}{R}}\; Claim(i,k,j)|\Xi,\; \overset{n|N}{\underset{i=1}{R}}\;\overset{F|N}{\underset{k=1}{R}}\; POS(i,k)|SM,$$

$$\overset{C|N}{\underset{i=1}{R}}\;\overset{F|N}{\underset{k=1}{R}}\; File(i)|S.Fact(k)|S>)$$

$\qquad\qquad)$

$\qquad)$

$\}$

Fig. 3. The algorithm of Parse|PM

The token ID refers to the index of the token in the sentence starting from 1. The POS tag is derived from one of the 17 universal POS tags known as ADJ (adjective), ADP (AD position), ADV (adverb), AUX (auxiliary verb), CONJ (coordinating conjunction), DET (determiner), INTJ (interjection), NOUN (noun), NUM (numeral), PART (particle), PRON (pronoun), PROPN (proper noun), PUNCT (punctuation), SCONJ (subordinating conjunction), SYM (symbol), VERB (verb) and X (other). The dependency refers to a syntactic relation between two tokens such as *nsubj* (subject), *dobj* (direct object), *pobj* (prepositional object), etc. It is noteworthy that ROOT is a special dependency indicator which is tagged to the root verb in Sentence|S. It does not have an associated token to pair with.

DSA|PM(<**I**:: $\overset{C|N}{\underset{i=1}{R}}$ *Claim*(*i*)|S>; <**O**:: $\overset{C|N}{\underset{i=1}{R}}$ *Sim*(*i*)|N ∈ {(0, Negative), (1, Positive), (2, Partial)}>;

$\qquad$ <**H**:: $\overset{n|N\ F|N}{\underset{i=1\ k=1}{R\ R}}$ *POS*(*i, k*)|SM, $\overset{C|N\ F_i|N\ 13}{\underset{i=1\ k=1\ j=1}{R\ R\ R}}$ *POS*(*i, k, j*)|Ξ, $\overset{C|N\ F_i|N}{\underset{i=1\ k=1}{R\ R}}$ *File*(*i*)|S.*Fact*(*k*)|S>)

$\overset{C|N\ F_i|N}{\underset{i=1\ k=1}{\{R\ R}}$ (// 1-F analysis

$\qquad \to F_i|N := |File(i)|S.Fact(k)|S|$

$\qquad$ // a) Parse claim and facts

$\qquad \to Claim|L := T|L$

$\qquad \rightarrowtail$ **Parse**|PM(<**I**:: *i*|N, *Claim*|L, *F_i*|N = 1, $\overset{F_i|N}{\underset{k=1}{R}}$ *Sentence*(*i,k*)|S>; <**O**:: $\overset{n|N\ F|N\ 13}{\underset{i=1\ k=1\ j=1}{R\ R\ R}}$ *POS*(*i, k, j*)|Ξ >;

$\qquad\qquad$ <**H**:: $\overset{n|N\ F|N}{\underset{i=1\ k=1}{R\ R}}$ *POS*(*i, k*)|SM, $\overset{C|N\ F_i|N}{\underset{i=1\ k=1}{R\ R}}$ *File*(*i*)|S.*Fact*(*k*)|S>)

$\qquad \rightarrowtail$ **StcComposition**|PM(<**I**:: $\overset{n|N\ F_i|N\ 13}{\underset{i=1\ k=1\ j=1}{R\ R\ R}}$ *POS*(*i, k, j*)|Ξ>; <**O**:: *Stc*|S>; <**H**:: *POS*|SM>)

$\qquad \to Claim|L := F|L$

$\qquad \to \overset{F|N}{\underset{k=1}{R}}$ *Sentence*(*i, k*)|S := $\overset{F|N}{\underset{k=1}{R}}$ *File*(*i*)|S.*Fact*(*k*)|S

$\qquad \rightarrowtail$ **Parse**|PM(<**I**:: *i*|N, *Claim*|L, *F_i*|N, $\overset{F_i|N}{\underset{k=1}{R}}$ *Sentence*(*i,k*)|S>; <**O**:: $\overset{n|N\ F_i|N\ 13}{\underset{i=1\ k=1\ j=1}{R\ R\ R}}$ *POS*(*i,k, j*)|Ξ >;

$\qquad\qquad$ <**H**:: $\overset{n|N\ F|N}{\underset{i=1\ k=1}{R\ R}}$ *POS*(*i, k*)|SM, $\overset{C|N\ F|N}{\underset{i=1\ k=1}{R\ R}}$ *File*(*i*)|S.*Fact*(*k*)|S>)

$\qquad \rightarrowtail$ **StcComposition**|PM(<**I**:: $\overset{n|N\ F_i|N\ 13}{\underset{i=1\ k=1\ j=1}{R\ R\ R}}$ *POS*(*i, k, j*)|Ξ>; <**O**:: *Stc*|S>; <**H**:: *POS*|SM>)

$\qquad$ // b) Differential semantic analysis (DSA)

$\qquad \rightarrowtail$ **DSA**|PM(<**I**:: *i*|N, *F_i*|N, *POS*(*i, k, j*)|Ξ>, <**O**:: $\overset{F|N}{\underset{k=1}{R}}$ *Match*(*i, k*)|N>;

$\qquad\qquad$ <**H**:: $\overset{C|N\ F|N}{\underset{i=1\ k=1}{R\ R}}$ *Diff*(*i, k*)|SM. $\overset{13}{\underset{j=1}{R}}$ *POS*(*i, k, j*)|Ξ; $\overset{F|N}{\underset{k=1}{R}}$ *File*(*k*)|S>)

$\qquad$)

$\qquad$ // c) Similarity determination: a claim vs. a paragraph

$\qquad \to \overset{F|N}{\underset{k=1}{R}}$ ($\to$ ◆*Match*(*i, k*)|N = 0

$\qquad\qquad \to Sim(i)|N := 0$

$\qquad\qquad \to \varnothing$

$\qquad\qquad \to$ ◆*Match*(*i, k*)|N = 1

$\qquad\qquad\qquad \to Sim(i)|N := 1$

$\qquad\qquad\qquad \to \varnothing$

$\qquad$)

$\qquad \to Sim(i)|N := 2$

)

}

Fig. 4. The algorithm of DSA

Parser|PM(<I:: *Sentence*|S>; <O:: $\displaystyle\mathop{R}_{i=1}^{n|N} \mathop{R}_{k=1}^{F_i|N} \mathop{R}_{j=1}^{13} POS(i,k,j)|\Xi$>;

        <H:: *DepTree*|SM, *POS*|SM>) ≜

{ ↣ Spacy|PM(<I:: *Sentence*|S>; <O:: *DepTree*|SM, #*Tokens*|N>); <H:: *DepTree*|SM>

// (a) Determin main POS

→ $\displaystyle\mathop{R}_{t=1}^{\#Token|N}$ (→ ((◆ *DepTree*(*t*)|SM.*Dep*|S = *ROOT*|S   // Verb detection

        ∧ *DepTree*(*t*)|SM.*Type*|S = *Verb*|S

        → *POS*(*i*, *k*, 2)|Ξ := *DepTree*(*t*)|SM.*Tokens*|S

        → *DepTree*(*t*)|SM.*Assoc*|N := *t*|N

)

    | (→ *a* := *DepTree*(*t*)|SM.*Assoc*|N

      → (◆ *DepTree*(*t*)|SM.*Dep*|S = *nsubj*|S   // Subject detection

        ∧ (*DepTree*(*t*)|SM.*Type*|S = *Noun*|S

         ∨ *DepTree*(*t*)|SM.*Type*|S = *Pron*|S

         ∨ *DepTree*(*t*)|SM.*Type*|S = *Propn*|S)

        ∧ (*DepTree*(*a*)|SM.*Dep*|S = *ROOT*|S)

        → *POS*(*i*, *k*, 1)|Ξ := *DepTree*(*t*)|SM.*Tokens*|S

        → *DepTree*(*t*)|SM.*Assoc*|N := *DepTree*(*a*)|SM.*Index*|N

)

)

    | (→ *a* := *DepTree*(*t*)|SM.*Assoc*|N

      → (◆ *DepTree*(*t*)|SM.*Dep*|S = *dobj*|S   // Object detection

        ∧ (*DepTree*(*t*)|SM.*Type*|S = *Noun*|S

         ∨ *DepTree*(*t*)|SM.*Type*|S = *Pron*|S

         ∨ *DepTree*(*t*)|SM.*Type*|S = *Propn*|S)

        ∧ (*DepTree*(*a*)|SM.*Dep*|S = *ROOT*|S)

        → *POS*(*i*, *k*, 3)|Ξ := *DepTree*(*t*)|SM.*Tokens*|S

        → *DepTree*(*t*)|SM.*Assoc*|N := *DepTree*(*a*)|SM.*Index*|N

)

)

)

)

}

Fig. 5. The algorithm of Parser|PM (part a)

The determination of the main POS components is made by three string comparisons shown in Fig. 5 between Dep|S and the dependency identifiers such as ROOT, *nsubj*, and *dobj* for verb, subject and object detection where the subject and object must be associated to the ROOT directly in order to avoid any misleading pairing. For example, sentence "*The dog eats a sausage*" will have a subject token pair {*dog*, *eats*} and an object

Parser|PM(<**I**:: *Sentence*|S>; <**O**:: $\underset{i=1}{\overset{n|N}{R}} \underset{k=1}{\overset{F_i|N}{R}} \underset{j=1}{\overset{13}{R}} POS(i, k, j)$|Ξ>;

    <**H**:: *DepTree*|SM, *POS*|SM>) ≙

{ ... // (b) Modifiers determination

 | (// MV detection

  → *v* := *DepTree*(*t*)|SM.*Assoc*|N

  → (◆ *DepTree*(*v*)|SM.*Dep*|S = ROOT|S

   → (◆*DepTree*(*t*)|SM.*Index*|N < *DepTree*(*t*)|SM.*Assoc*|N

    (◆ (*DepTree*(*t*)|SM.*Dep*|S = aux|S ∨ neg|S) // Identify MV1

     → *POS*(*i*, *k*, 10)|Ξ := *DepTree*(*t*)|SM.*Token*|S

    | ∼*DepTree*(*t*)|SM.*Dep*|S = advmod|S // MV1

     → *POS*(*i*, *k*, 10)|Ξ := *DepTree*(*t*)|SM.*Token*|S

)

   | ◆*DepTree*(*t*)|SM.*Index*|N > *DepTree*(*t*)|SM.*Assoc*|N ∧

    (◆*DepTree*(*t*)|SM.*Dep*|S = neg|S // Identify MV2

     → *POS*(*i*, *k*, 11)|Ξ := *DepTree*(*t*)|SM.*Tokens*|S

    | ∼*DepTree*(*t*)|SM.*Dep*|S = advmod|S // MV2

     → *POS*(*i*, *k*, 10)|Ξ := *DepTree*(*t*)|SM.*Token*|S

)

)

)

 | (// MJ detection

  → *j* := *DepTree*(*t*)|SM.*Assoc*|N

  → ◆ (*DepTree*(*j*)|SM.*Dep*|S ≠ ROOT|S

   → (◆ (*POS*(*i*, *k*, 1)|Ξ ≠ " "

    → (◆*DepTree*(*t*)|SM.*Index*|N < *DepTree*(*j*)|SM.*Index*|N // MJ1

     → *POS*(*i*, *k*, 8)|Ξ := *POS*(*i*, *k*, 8)|Ξ + *DepTree*(*t*)|SM.*Token*|S

    | ◆∼ // MJ2

     → *POS*(*i*, *k*, 9)|Ξ := *POS*(*i*, *k*, 9)|Ξ + *DepTree*(*t*)|SM.*Token*|S

)

)

  | ◆*DepTree*(*t*)|SM.*Dep*|S = advmod|S // MJ2

    → *POS*(*i*, *k*, 9)|Ξ := *POS*(*i*, *k*, 9)|Ξ + *DepTree*(*t*)|SM.*Token*|S

)

)

 | (// MO detection

  → *o* := *DepTree*(*t*)|SM.*Assoc*|N

  → ◆ (*DepTree*(*o*)|SM.*Dep*|S ≠ ROOT|S

   → (◆ (*POS*(*i*, *k*, 3)|Ξ ≠ " "

    → (◆*DepTree*(*t*)|SM.*Index*|N < *DepTree*(*o*)|SM.*Index*|N ∧

     DepTree(*t*)|SM.*Index*|N > *DepTree*(*v*)|SM.*Index*|N

     // MO1

     → *POS*(*i*, *k*, 12)|Ξ := *POS*(*i*, *k*, 12)|Ξ + *DepTree*(*t*)|SM.*Token*|S

    | ◆*DepTree*(*t*)|SM.*Index*|N > *DepTree*(*o*)|SM.*Index*|N ∧

     DepTree(*t*)|SM.*Index*|N > *DepTree*(*v*)|SM.*Index*|N

     // MO2

     → *POS*(*i*, *k*, 13)|Ξ := *POS*(*i*, *k*, 13)|Ξ + *DepTree*(*t*)|SM.*Token*|S

)

   | ◆*DepTree*(*t*)|SM.*Dep*|S = advmod|S // MO2

    → *POS*(*i*, *k*, 13)|Ξ := *POS*(*i*, *k*, 13)|Ξ + *DepTree*(*t*)|SM.*Token*|S

)

)

)

}

Fig. 6. The algorithm of Parser|PM (part b)

token pair {*eats, sausage*} given "*eats*" as the root verb. Figure 6 illustrates complicated situations for eliciting modifiers for verb, subject and object using the DepTree|SM structure with certain identifiers. The pre- and post-modifiers are determined based on their positions in the sentence according to the main POS components.

Figure 7 describes StcComposition|PM for extracting time and space variants in DepTree|SM. Those special modifiers may appear as clauses or prepositional phrases in a sentence. The space/time clauses are determined by the token identifiers *where* and *when*. A prepositional phrase will be extracted when a special preposition is detected to be associated to its noun objects (i.e.: *for 5 min*, *via email*, etc.). There are some special prepositions such as *at* or *in* that are widely used in both space and time phrases (i.e.: *at 5 pm, at school, in the room*). In this case, a classifier is attached to distinguish them based on its space/time properties. After the clauses and phrases are elicited, the POS structure is updated to its final state. The completed POS will be adopted by DSA|PM as the parsing analysis results in the following section.

3.2 Algorithm of Differential Semantic Analysis (DSA)

The algorithm of *Differential Semantic Analysis* (DSA) is described in Process (b) of Fig. 8, which recursively carries out paragraph learning based on the Claim|SM and Fact|SM refined in Parse|PM. The semantic determination of fake news in the DSA|PM algorithm is aggregated from the analysis of their syntactic matches, semantic consistency, and supplemented by a macro statistical score learnt at the levels of concepts, sentences, and paragraphs from the bottom up.

The DSA|PM shown in Fig. 8 performs a 1-to-n differential analysis in order to determine Diff|Ξ for each of the POS components as well as the aggregated sentence structure Stc|Ξ as defined bellow.

Definition 5. The *differential matching score* Diff|Ξ of a fake news Claim|S is determined by its level of consistency with respect to known facts (Fact|S):

$$Diff \mid \Xi \triangleq FN \mid S = Claim \mid S \oplus Fact \mid S$$
$$= \frac{\mid Claim \mid \Xi \cap Fact \mid \Xi \mid}{\mid Claim \mid \Xi \cup Fact \mid \Xi \mid} \tag{4}$$

According to Definition 5, a pair of POS components for a claim and associated facts is considered as semantically identical if Diff|Ξ results in an empty set ∅. However, in case of synonyms involved, a lookup table is adopted based on Python Dictionary. That is, synonyms are treated as identical in DSA|PM. In case that multiple modifiers are used to describe a certain subject, verb or object, they will be considered as coherent if they share the same key identifiers. Those identifiers can be nouns, adjectives, adverbs or phrases that refine the identities for the main POS components. A pair of coherent modifiers is also treated as no difference (Diff|Ξ = ∅) in the DAS|PM algorithm. Furthermore, because the modifiers and space/time parameters may be absent in a sentence, they will not influence the differential analyses and will be marked as nonapplicable, so that they are exempted during the determination of the matching score.

SentenceComposition|PM(<I:: *DepTree*|SM>; <O:: *Stc*|S>; <H:: $\overset{n|N}{\underset{i=1}{R}} \overset{F_i|N}{\underset{k=1}{R}} \overset{13}{\underset{j=1}{R}} POS(i,k,j)|\Xi$>) ≜

{ // a) Space/Time Clause determination

     ⟩ *isClause* := F|L

→ $\overset{\#Token|N}{\underset{t=1}{R}}$ (// Clause

         → *word* := *DepTree*(n)|SM.*Token*|S

         → ◆ (*word*.lower() = "when"

             → APPEND(*POS*(i, k, 5)|Ξ, *POS*(i, k, j)|Ξ[n : end])

             → *isClause* := T|L

)

         → ◆ (*word*.lower() = "where"

             → APPEND(*POS*(i, k, 4)|Ξ, *POS*(i, k, j)|Ξ[n : end])

             → *isClause* := T|L

)

)

→ (◆ *isClause* = F|L // Prepositional phrase

     → *isTime*|L := F|L

     → $\overset{\#Token|N}{\underset{t=1}{R}}$ (→ *dep* := *DepTree*(n)|SM.*dependency*|S

         → ◆ *dep* = "pobj" // ie: "for 5 minutes"

             → *range* := (*DepTree*(n)|SM.*Assoc*|N, *DepTree*(n)|SM.*ID*|N)

)

     ↦ ExtractText (<I:: *range*, *DepTree*(n)|SM>, <O:: *p_phrase*|Ξ>)

     → (◆ ∃*word*|S ∈ *p_phrase*|Ξ ∧ *word*|S ∈ *TimeKeyword*|Ξ

         → *isTime*|L := T|L

         → APPEND(*POS*(i, k, 5)|Ξ, *p_phrase*|Ξ) // Time phrase

)

     → (◆ *isTime*|L = F|L ∧ (*p_phrase*|Ξ(1) = "in" ∨

     p_phrase|Ξ(1) = "at" ∨ *p_phrase*|Ξ(1) = "via"

         → APPEND(*POS*(i, k, 4)|Ξ, *p_phrase*|Ξ) // Space phrase

)

)

// b) Update PC/TC/XC and compose Stc

→ *POS*(i, k, 4)|SM.*PC*|S := *POS*(i, k, 4)|Ξ ∈ *SPW*|Ξ

→ *POS*(i, k, 5)|SM.*TC*|S := *POS*(i, k, 5)|Ξ ∈ *TPW*|Ξ

→ *POS*(i, k, 6)|SM.*XC*|S := *POS*(i, k, 6)|Ξ ∉ *SPW*|Ξ ∪ *TPW*|Ξ

→ *POS*(i, k, 7)SM.*Stc*|S := *POS*(i, k, 8)|S + *POS*(i, k, 1)|S + *POS*(i, k, 9)|S + // MJ1, JP, MJ2

                       POS(i, k, 10)|S + *POS*(i, k, 2)|S + *POS*(i, k, 11)|S + // MV1, VP, MV2

                       POS(i, k, 12)|S + *POS*(i, k, 3)|S + *POS*(i, k, 13)|S + // MO1, OP, MO2

                       POS(i, k, 4)|S + *POS*(i, k, 5)|S + *POS*(i, k, 6)|S // PC, TC, XC

}

Fig. 7. The algorithm of SentenceComposition|PM

After the set of Diff|Ξ assessments are obtained, the fact sentences with different subjects to the claim will be eliminated to optimize the analyses results. The determination of negative claims is then followed by initializing Match(i, k)|N to 2. A fact sentence is negative if any of the NegW|Ξ keywords is detected that indicates a contradiction to the claim. The matching score will be set to 0 for the first negative occurrence, while it remains 2 otherwise.

The determination of positive similarity starts if the negative test has been ruled out. The positive verification is done through either a semantic or a statistical method. A fact sentence is semantically positive if the Diff|Ξ values of the sets of main POS components, the space and time variant, and one of the pre/post modifiers are ∅. It may also be statistically positive if the differential Stc|Ξ is below a certain threshold θ.

DSA|PM finally determines a set of matching scores $\overset{F|N}{\underset{k=1}{R}} Match(i, k)|N$ for each ith claim against all F_i|N number of facts. The assessment scores are iteratively calculated based on the above differential analysis algorithm, which will be applied for determining the final similarity in the following subsection.

3.3 Determination of Fake News by Semantic Inconsistency Assessment

According to Definition 5, the *semantic consistency* δ of a piece of news is determined by the *extent of similarity* between a certain piece of news and its background facts that it is based. That is, δ|I = 1 − Diff|I. In Process (c) of Algorithm DSA|PM, a final similarity score δ|I is derived based on the set of semantic matching assessments carried out in Process (b).

δ|N denotes the outcome of the DSA|PM algorithm where three assessment results may be obtained:

a) δ|N = 0: It indicates an inconsistency between the claim and one or more facts so the news is classified as fraudulent.
b) δ|N = 1: It denotes that the claim is completely fulfilled with or supported by the facts. Then, it is recognized as a true news.
c) 0 < δ|N < 1: It implies that the truthiness of the claim falls between Categories (a) and (b). In this case, the given claim lacks critical facts to either support or against the facts. Hence, the claim is classified as partially true.

As shown in Fig. 8, the determination of the negative similarity is carried out first. If a fact is found to be contradiction to the claim, Similarity will be set to δ|N - 0 and DSA will move on to the next claim immediately. However, a non-negative claim will then be tested against its positiveness. If the claim is true or possesses critical supporting fact(s) found in the given facts, the claim will be confirmed with δ|N = 1. Otherwise, DSA will assign δ|N = 2 (partially true) if both previous tests are failed.

$$\mathbf{DSA|PM}(<\mathbf{I}::\ i|\mathrm{N},\ F_i|\mathrm{N},\ \mathit{Claim}|\mathrm{SM},\ \mathit{Fact}|\mathrm{SM}>,\ <\mathbf{O}::\ \overset{F_i\mathrm{N}}{\underset{k=1}{R}}\ \mathit{Match}(i,k)|\mathrm{N}>;$$

$$<\mathbf{H}::\ \overset{C\mathrm{N}\ F_i\mathrm{N}}{\underset{i=1\ k=1}{R\ R}}\ \mathit{Diff}(i,k)|\mathrm{SM},\ \overset{F_i\mathrm{N}}{\underset{k=1}{R}}\ \mathit{File}(k)|\mathrm{S}>$$

$$\{//\ \mathit{Diff}|\Xi = \oplus\ := (\mathit{Claim}|\Xi \cap \mathit{Fact}|\Xi) \setminus (\mathit{Claim}|\Xi \cup \mathit{Fact}|\Xi)$$

$$\to\ \overset{F_i\mathrm{N}}{\underset{k=1}{R}}\quad (//\ \text{a) 1-F differential analyses}$$

// Subject
$$\to\ \mathit{Diff}(i,k)|\mathrm{SM}.JP|\Xi := \mathit{Claim}(i)|\mathrm{SM}.JP|\Xi \oplus \mathit{Fact}(i,k)|\mathrm{SM}.JP|\Xi$$

// Verb
$$\to\ \mathit{Diff}(i,k)|\mathrm{SM}.VP|\Xi := \mathit{Claim}(i)|\mathrm{SM}.VP|\Xi \oplus \mathit{Fact}(i,k)|\mathrm{SM}.VP|\Xi$$

// Object
$$\to\ \mathit{Diff}(i,k)|\mathrm{SM}.OP|\Xi := \mathit{Claim}(i)|\mathrm{SM}.OP|\Xi \oplus \mathit{Fact}(i,k)|\mathrm{SM}.OP|\Xi$$

// Pre/post subject modifier
$$\to\ \mathit{Diff}(i,k)|\mathrm{SM}.MP1|\Xi := \mathit{Claim}(i)|\mathrm{SM}.MP1|\Xi \oplus \mathit{Fact}(i,k)|\mathrm{SM}.MP1|\Xi$$
$$\to\ \mathit{Diff}(i,k)|\mathrm{SM}.MP2|\Xi := \mathit{Claim}(i)|\mathrm{SM}.MP2|\Xi \oplus \mathit{Fact}(i,k)|\mathrm{SM}.MP2|\Xi$$

// Pre/post verb modifier
$$\to\ \mathit{Diff}(i,k)|\mathrm{SM}.MV1|\Xi := \mathit{Claim}(i)|\mathrm{SM}.MV1|\Xi \oplus \mathit{Fact}(i,k)|\mathrm{SM}.MV1|\Xi$$
$$\to\ \mathit{Diff}(i,k)|\mathrm{SM}.MV2|\Xi := \mathit{Claim}(i)|\mathrm{SM}.MV2|\Xi \oplus \mathit{Fact}(i,k)|\mathrm{SM}.MV2|\Xi$$

// Pre/post object modifier
$$\to\ \mathit{Diff}(i,k)|\mathrm{SM}.MO1|\Xi := \mathit{Claim}(i)|\mathrm{SM}.MO1|\Xi \oplus \mathit{Fact}(i,k)|\mathrm{SM}.MO1|\Xi$$
$$\to\ \mathit{Diff}(i,k)|\mathrm{SM}.MO2|\Xi := \mathit{Claim}(i)|\mathrm{SM}.MO2|\Xi \oplus \mathit{Fact}(i,k)|\mathrm{SM}.MO2|\Xi$$

// Space/time modifier
$$\to\ \mathit{Diff}(i,k)|\mathrm{SM}.PC|\Xi := \mathit{Claim}(i)|\mathrm{SM}.PC|\Xi \oplus \mathit{Fact}(i,k)|\mathrm{SM}.PC|\Xi$$
$$\to\ \mathit{Diff}(i,k)|\mathrm{SM}.TC|\Xi := \mathit{Claim}(i)|\mathrm{SM}.TC|\Xi \oplus \mathit{Fact}(i,k)|\mathrm{SM}.TC|\Xi$$
$$\to\ \mathit{Diff}(i,k)|\mathrm{SM}.XC|\Xi := \mathit{Claim}(i)|\mathrm{SM}.XC|\Xi \oplus \mathit{Fact}(i,k)|\mathrm{SM}.XC|\Xi$$

// Setence
$$\to\ \mathit{Diff}(i,k)|\mathrm{SM}.Stc|\Xi := \mathit{Diff}(i,k)|\mathrm{SM}.(JP|\Xi \cup VP|\Xi \cup OP|\Xi \cup$$
$$MP1|\Xi \cup MP2|\Xi \cup$$
$$MV1|\Xi \cup MV2|\Xi \cup$$
$$MO1|\Xi \cup MO2|\Xi \cup$$
$$PC|\Xi \cup TC|\Xi \cup XC|\Xi)$$

// b) Eliminate irrelevant sentence
$$\to\ \blacklozenge\ \mathit{Diff}(i,k)|\mathrm{SM}.JP|\Xi \neq \varnothing$$
$$\to\ \varnothing$$

// c) Negative sentence determination
$$\to\ \mathit{Match}(i,k)|\mathrm{N} := 2$$
$$\to (\ \blacklozenge\ \mathit{Diff}(i,k)|\mathrm{SM}.Stc|\Xi \in NegW|\Xi$$
// $NegW|\Xi$ = {no, not, never, few, hardly, scarcely, barely, narrowly, rarely,
                       little, seldom, seldomly, false, rumor, fraud, fake,...}
$$\to\ \mathit{Match}(i,k)|\mathrm{N} := 0$$
)

// d) Positive similarity determination
$$\to\ \blacklozenge\ \mathit{Match}(i,k)|\mathrm{N} \neq 0$$
// Semantic positive
$$\to (\ \blacklozenge\ \mathit{Diff}(i,k)|\mathrm{SM}.JP|\Xi = \varnothing \wedge \mathit{Diff}(i,k)|\mathrm{SM}.VP|\Xi = \varnothing \wedge$$
$$\mathit{Diff}(i,k)|\mathrm{SM}.OP|\Xi = \varnothing \wedge$$
$$(\mathit{Diff}(i,k)|\mathrm{SM}.MP1|\Xi = \varnothing \vee \mathit{Diff}(i,k)|\mathrm{SM}.MP2|\Xi = \varnothing) \wedge$$
$$(\mathit{Diff}(i,k)|\mathrm{SM}.MV1|\Xi = \varnothing \vee \mathit{Diff}(i,k)|\mathrm{SM}.MV2|\Xi = \varnothing) \wedge$$
$$(\mathit{Diff}(i,k)|\mathrm{SM}.MO1|\Xi = \varnothing \vee \mathit{Diff}(i,k)|\mathrm{SM}.MO2|\Xi = \varnothing) \wedge$$
$$\mathit{Diff}(i,k)|\mathrm{SM}.PC|\Xi = \varnothing\ \wedge \mathit{Diff}(i,k)|\mathrm{SM}.TC|\Xi = \varnothing)$$
$$\to\ \mathit{Match}(i,k)|\mathrm{N} := 1$$
// Statistical positive
$$|\blacklozenge\ |\mathit{Diff}(i,k)|\mathrm{SM}.Stc|\Xi|\ < \theta$$
// Calibrate θ
// $\theta = 0.2 * | (\mathit{Claim}(i)|\mathrm{SM}.Stc|\Xi \cup \mathit{Fact}(i,k)|\mathrm{SM}.Stc|\Xi)|$
$$\to\ \mathit{Match}(i,k)|\mathrm{N} := 1$$
)
)
}

Fig. 8. The algorithm of DSA|PM

4   Implementation and Experimental Results of AFNR

Based on the DSA|PM algorithm as formally designed in the previous section, the AFNR system for autonomous fake news recognitions is implemented and verified by a large set of experimental case studies. Representative samples in different categories of true, fake, or partially true news are demonstrated by rigorous differential syntax and semantic analyses. Then, large scale and random experiments are carried out by AFNR for testing its overall performance beyond traditional data-driven neural network technologies for fake news recognitions.

4.1   System Implementation and Environment of AFNR

The AFNR system is implemented and validated in an integrated Python developing environment known as Anaconda [27]. Anaconda is a multi-platform package management tool that is designed for scientific data processing with a coherent software engineering environment. It is featured by a rich third-party library as well as strong and consistent developer community support. In Anaconda, projects are independent across different environments created by users. This framework has simplified the system configuration for variant projects with different software requirements and prevents cross interference among the programs. Therefore, Anaconda is adopted as a powerful tool for solving dependency issues when a large set of software and data packages are involved. This is particularly critical to this project that requires 10 + third-party libraries to run coherently including the Natural Language Tool Kit (NLTK) [28] and Spacy [26]. In addition, Anaconda supports both Windows and Linux systems, which enables the execution and testing of the DSA algorithm in a local machine or a remote virtual system.

In the AFNR system, the Amazon virtual machine has been adopted to enable team collaboration, code management and remote access. EC2 is equipped with Ubuntu 16.04 LTS as the operating system. This integrated Linux system has Python and Python 3 pre-installed. It provides sufficient computing hardware configuration for supporting the AFNR system. Contrary to a neural network program that requires lengthy training, AFNR is training free and running at a performance level of approximately 20 claims per minute on the EC2 virtual machine.

Once an EC2 instance is launched in the system, it uses WinSCP [29] and Putty [26] to integrate the local and remote machines. WinSCP provides an intuitive window-by-window user interface for file transferring. Files in different locations can be easily transferred by drag-and-drop across the Windows and Putty environments. Though, users are allowed to run certain program via a remote Python editor, it is recommended to upload any remote code to the VM via WinSCP in order to avoid network communication lags and any possible discrepancies.

4.2   Sample Fake News Analyses

In this session, three representative sample news items with known facts have been selected from the DataCup'19 database [23]. They are analyzed and verified by the AFNR system driven by the DSA algorithm to demonstrate the functions and their performances.

> **Claim 1**: "Maine legislature candidate Leslie Gibson insulted Parkland shooting survivor and activist Emma Gonzalez via Twitter."
>
> **Fact 1.1**: "Maine House candidate calls Parkland student activist Emma Gonzalez skinhead lesbian."

Fig. 9. Test case 1

The first sample shown in Fig. 9 is a typical example in the dataset where the POS components of the fact nearly fulfill the claim. For instance, they share the same subject *"candidate"* and object *"Emma Gonzalez"*, but the verbs and some modifiers differ. According to Definition 5, the preliminary differential analysis is obtained as shown in Eq. 5.

$$Diff_1|\Xi \triangleq FN|S = Claim_1|S \oplus Fact_{1.1}|S$$

$$= \begin{cases} JP : \varnothing \\ VP : \text{insult, call} \\ OP : \varnothing \\ MJ1 : \text{House, legislature,} \\ MJ2 : \text{Leslie Gibson} \\ MV : \varnothing \\ MO1 : \text{student, shooting survivor} \\ MO2 : \text{skinhead lesbian} \\ SC : \text{Twitter} \\ TC : \varnothing \end{cases} \tag{5}$$

According to the DSA algorithm, a claim is considered as true (Sim|N = 1) if the differentiation result are empty sets across all POS components, unless the outstanding ones are semantically coherent, and therefore, can be treated as identical. Claim 1 is determined as true because the verb *"insult"* has the same meaning as *"call [someone] skinhead lesbian"*, where *"call"* and *"skinhead lesbian"* are the verb and post-object modifier detected in the fact. The pre-subject modifiers for both sentences are also considered identical because word *"House"* is a synonym of *"legislature"*. These interpretation rules have been set up in a DSA synonym dictionary during the program initialization and may be extended when it is needed by the system. Furthermore, the pre-object modifiers for the claim and fact are coherent as they share *"Parkland"* and *"activist"* as key identifiers for *"Emma Gonzalez"*. Finally, *"Twitter"* in the claim is omitted in this case since the fact lacks such space variant. According to DSA|PM in Sect. 3.2, Claim 1 is judged as true news because of: a) Identical JP, VP and OP; b) Coherent pre/post modifiers; and c) Inapplicable space variant.

The second claim given in Fig. 10 is an example of fake news due to the contradiction between the claim and fact where the negation keywords *"false"*, and *"rumor"* are predefined in NegW|Ξ.

> **Claim 2:** "Koch Industries paid the legal fees of George Zimmerman."
>
> **Fact 2.1:** "Koch Industries have also stated that the statement is a completely false and baseless rumor."

Fig. 10. Test case 2

Claim 3 shown in Fig. 11 is recognized as partially true (Sim|N = 2) after it fails both the negative and positive tests. Although the facts are relevant to the claim since they both describe poverty state of "*Appalachia*," they neither contradict nor positive in corresponding POS components. Therefore, the claim is classified as partially true.

> **Claim 3:** "Appalachia is the poorest country in the U.S. and happens to be more than 90 percent white."
>
> **Fact 3.1:** "Appalachia had its poverty rate, 31 percent in 1960 and was 16.7 percent over the 2012–2016 period."
>
> **Fact 3.2:** "Central Appalachia in particular still battles economic distress, with concentrated areas of high poverty"

Fig. 11. Test case 3

4.3 Experimental Results on Large-scale Samples

The AFNR system and the DSA algorithm have been applied to a large dataset of 876 samples [23]. Two testing strategies have been designed in the experiments for both black box and white box tests. In the black box approach, the entire sample dataset is classified to three categories: positive, negative, and partially true. Table 1 shows the result of accuracy for each category obtained by the AFNR system.

Table 1. Testing results in data-oriented approach

Claim Type	Total	Identified	Accuracy (%)
Negative	425	297	69.88
Positive	103	72	69.90
Partial	348	245	70.40
Overall	876	614	70.01

Table 1 is derived based on the testing result of each claim and the assessment results in one of three categories (Positive, Negative, Partial). An overall accuracy of 70.01% among 876 samples has been obtained. This black box approach is useful to classify the testing results in order to identify any failure case instantly. It also provides an overview of the algorithm on each partitioned category.

In the white box approach, we applied DSA|PM as three determination functions to deal with separated cases of Sim|N = [0, 1, 2]. Every function is executed independently and returns true if the similarity of the tested claim matches the tool's expectation. The result shown in Table 2 indicates that among 425 negative samples, 297 are correctly determined but 29 of them are mis-categorized to positive (Sim|N = 1) and 99 samples are falsely detected as partial (Sim|N = 2). The bold numbers in each category denote the claims that are successfully verified in Table 2.

Table 2. Testing Results in the Function-Oriented Approach

Category	Negative	Positive	Partial
Sim|N = 0	**297**	8	30
Sim|N = 1	29	**72**	73
Sim|N = 2	99	23	**245**
Total	425	103	348
Accuracy (%)	69.88	69.90	70.40

Figure 12 shows the partitions in percentage with the number of tested claims labelled on the bars. Compared with the first method of ATNR applications, we divide the algorithm into sub-functions for revealing insights of the algorithm in order to identify logical issues in an intuitive way. As a result, these two approaches have obtained identical accuracy results that prove the consistency of the DSA theory and the AFNR methodology.

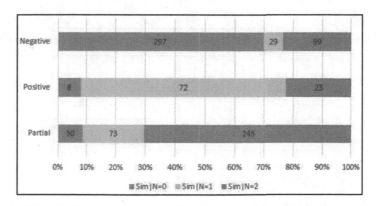

Fig. 12. Summary of the experimental results by the AFNR system

4.4 Comparative Analyses between the Autonomous AFNR and Neural-Network-Based Technologies

The experimental results in Sect. 4.3 demonstrate that the DSA algorithm and its underpinned theories have enabled efficient fake news recognition rigorously and autonomously. A comparative analysis between the AFNR and DataCup'19 results is illustrated in Fig. 13. It is observed that, in DataCup'19 [23], the top three participants are ranked as 55%, 53% and 51%, respectively, by adopting conventional neural network-based algorithms. The overall average accuracy rate of the top 100 teams has only reached 20.5%, because the majorities were below 30%. However, considering the entire 500 + participant teams, the average benchmark could be lower than 10.0%.

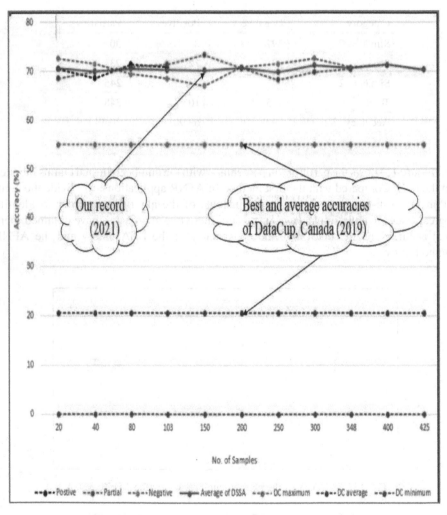

Fig. 13. Testing results generated by the AFNR system

However, the AFNR system powered by the DSA algorithm has shown an outstanding accuracy of 70.01%. Compared to the neural network technology used by the DataCup'19 teams, the autonomous AFNR system has taken advantages enabled by the following features: a) AFNR is an autonomous methodology that is free of training underpinned by rigorous software science theories [32–34] and intelligent mathematics [1]; b) Running the DSA algorithm may only require minimum hardware support as described in Sect. 4.1; and c) AFNR is implemented in Python and completely transparent to developers and users. This important feature allows white-box testing on the logic segments in the program for fast and precise debugging and functional improvements.

5 Conclusion

It has been recognized that fake news detection and recognitions are a fundamental challenge to AI, cognitive computing, and machinable semantic comprehension. This work has presented a formal theory of fake news recognition by cognitive semantic computing powered by contemporary Intelligent Mathematics (IM) known as Concept Algebra and Semantic Algebra. This work has designed and implemented the *Autonomous Fake News Recognition* (AFNR) system based on a training-free methodology for *Differential Sematic Analysis* (DSA). AFNR is supported by the Anaconda, MATLAB, and python environment with the Natural Language Toolkit (NLTK) and the English parser (Spacy). The DSA algorithm has demonstrated a significant improvement for fake news recognition, which has out-performed traditional data-driven machine learning technologies based on a large-scale fake news database. The AFNR system for fake news recognition has enabled autonomous semantic comprehension of natural language expressions with an outstanding processing accuracy of 70.1% beyond the best reported result of 55.0% in DataCup'19.

Acknowledgement. This work is supported in part by the DND IDEaS program, the IEEE Autonomous System Initiative (ASI), IEEE SMC Society Technical Committee on Brain-inspired Cognitive Systems (TC-BCS), the IEEE Computer Science Society Technical Committee on Computational Life Sciences (TC-CLS), and the Intelligent Mathematics (IM) initiative of the International Institute of Cognitive Informatics and Cognitive Computing (I2CICC).

Conflict of Interest Statement. The authors declare that the research was conducted in the absence of any commercial or financial relationships that could be construed as a potential conflict of interest.

References

1. Wang, Y., et al.: An odyssey towards the next generation of computers: theoretical discoveries, intelligent mathematics, and technological breakthroughs of cognitive computers. Fron. Comput. Sci. **5**, 1–17 (2023)
2. Castillo, C., Mendoza, M., Poblete, B.: Information credibility on twitter. In: The 20th International Conference on World Wide Web, pp. 675–684. ACM (2011)

3. Ferreira, W., Vlachos, A.: Emergent: a novel data-set for stance classification. In: Conference of the North American Chapter of Association for Computational Linguistics: Human Language Technologies, pp. 1163–1168. ACL (2016)
4. Gupta, A., Kumaraguru, P., Castillo, C., Meier, P.: Tweetcred: real-time credibility assessment of content on twitter. In: Aiello, L.M., McFarland, D. (eds.) SocInfo 2014. LNCS, vol. 8851, pp. 228–243. Springer, Cham (2014). https://doi.org/10.1007/978-3-319-13734-6_16
5. Kwon, S., Cha, M., Jung, K.: Rumor detection over varying time windows. PLoS ONE **12**(1), 0168344 (2017)
6. Ma, J., et al.: Detecting rumors from microblogs with recurrent neural networks. In: IJCAI, pp. 3818–3824 (2016)
7. Ma, J., Gao, W., Wei, Z., Lu, Y., Wong, K.: Detect Rumors using time series of social context information on microblogging websites. In: 24th ACM International Conference on Information and Knowledge Management, pp. 1751–1754. ACM (2015)
8. Markines, B., Cattuto, C., Menczer, F.: Social spam detection. In: 5th International Workshop on Adversarial Information Retrieval on the Web, pp. 41–48. ACM (2009)
9. Markowitz, D., Hancock, J.: Linguistic traces of a scientific fraud: the case of Diederik Stapel. PLoS ONE **9**(8), e105937 (2014)
10. Rubin, V., Chen, Y., Conroy, N.: Deception detection for news: three types of fakes. Proc. Assoc. Inf. Sci. Technol. **52**(1), 1–4 (2015)
11. Shu, K., Sliva, A., Wang, S., Tang, J., Liu, H.: Fake news detection on social media: a data mining perspective. ACM SIGKDD Expl. Newsl **19**, 22–36 (2017)
12. Pan, J.Z., Pavlova, S., Li, C., Li, N., Li, Y., Liu, J.: Content based fake news detection using knowledge graphs. In: Vrandečić, D., et al. (eds.) ISWC 2018. LNCS, vol. 11136, pp. 669–683. Springer, Cham (2018). https://doi.org/10.1007/978-3-030-00671-6_39
13. Wu, K., Yang, S., Zhu, K.: False rumors detection on sina weibo by propagation structures. In: IEEE 31st International Conference on Data Engineering (ICDE), pp. 651–662. IEEE (2015)
14. Zhao, Z., Resnick, P., Mei, Q.: Enquiring minds: early detection of rumors in social media from enquiry posts. In: 24th International Conference on World Wide Web, pp. 1395–1405. ACM (2015)
15. Wang, Y.: Towards the synergy of cognitive informatics, neural informatics, brain informatics, and cognitive computing. Int. J. Cogn. Inform. Nat. Intell. **5**(1), 75–93 (2011)
16. Wang, Y.: A formal syntax of natural languages and the deductive grammar. Fund. Inform. **90**(4), 353–368 (2009)
17. Wang, Y.: Concept algebra: a denotational mathematics for formal knowledge representation and cognitive robot learning. J. Adv. Math. Appl. **4**(1), 62–87 (2015)
18. Wang, Y.: On semantic algebra: a denotational mathematics for cognitive linguistics, machine learning, and cognitive computing. J. Adv. Math. Appl. **2**(2), 145–161 (2013)
19. Wang, Y., et al.: On the philosophical, cognitive and mathematical foundations of symbiotic autonomous systems. Phil. Trans. R. Soc. A **379**, 20200362, pp.1–20. https://doi.org/10.1098/rsta.2020.0362
20. Wang, Y., Berwick, R.C.: Formal relational rules of English syntax for cognitive linguistics, machine learning, and cognitive computing. J. Adv. Math. Appl. **2**(2), 182–195 (2013)
21. Wang, Y., et al.: Towards a theoretical framework of autonomous systems underpinned by intelligence and systems sciences. IEEE/CAA J. Autom. Sinica **8**(1), 52–63 (2021)
22. Wang, Y.: A formal theory of AI trustworthiness for evaluating autonomous AI systems. In: IEEE 2022 International Conference on System, Man, and Cybernetics (SMC 2022), Prague, Czech, pp. 137–142 (2022)
23. DataCup (2019). https://www.datacup.ca/
24. Wang, Y.: From data-aggregative learning to cognitive knowledge learning enabled by autonomous AI theories and intelligent mathematics (Keynote). In: 2022 Future Technologies Conference (FTC 2022), Vancouver, Canada, October, pp. 1–3 (2022)

25. Wang, Y.: The real-time process algebra (RTPA). Ann. Software Eng. **14**, 235–274 (2002)
26. Spacy project. Industrial-Strength Natural Language Processing in Python (2021). https://spacy.io/
27. Anaconda (2021). https://www.anaconda.com/
28. NLTK project. Natural Language Toolkit (2020). https://www.nltk.org/install.html
29. WinSCP (2021). WinSCP. https://winscp.net/eng/index.php
30. Stanford Univ. Stanford Parser, The Stanford Natural Language Processing Group (2019). http://nlp.stanford.edu:8080/parser/index.jsp.
31. Wang, Y.: Software Engineering Foundations: A Software Science Perspective. Auerbach Publications, New York (2007)
32. Wang, Y.: Mathematical laws of software. Trans. Comput. Sci. **2**, 46–83 (2008)
33. Wang, Y.: On cognitive computing. Int. J. Software Sci. Comput. Intell. **1**(3), 1–15 (2009). https://doi.org/10.4018/jssci.2009070101
34. Wang, Y.: Software science: on general mathematical models and formal properties of software. J. Adv. Math. Appl. **3**(2), 130–147 (2014)
35. Wang, Y.: On formal and cognitive semantics for semantic computing. Int. J. Semantic Comput. **4**(2), 203–237 (2010)
36. Wang, Y., et al.: Brain-inspired systems: a transdisciplinary exploration on cognitive cybernetics, humanity, and systems science toward autonomous artificial intelligence. IEEE Syst. Man Cybern. Mag. **6**(1), 6–13 (2020)
37. Wang, Y.: On cognitive informatics. Brain Mind Transdisciplinary J. Neurosci. Neurophilosophy **4**(2), 151–167 (2003)
38. Wang, Y.: The theoretical framework of cognitive informatics. Int. J. Cogn. Inform. Natural Intell. **1**(1), 1–27 (2007)
39. Wang, Y., Berwick, R.C., Haykin, S., Pedrycz, W., Kinsner, W., et al.: Cognitive informatics and cognitive computing in year 10 and beyond. Int. J. Cogn. Inf. Natural Intell. **5**(4), 1–2 (2011)
40. Wang, Y., et al.: A survey and formal analyses on sequence learning methodologies and deep neural networks. In: 17th IEEE International Conference on Cognitive Informatics and Cognitive Computing (ICCI*CC'18), University of California, Berkeley, USA, July 16–18, pp. 6–15. IEEE CS Press (2018)
41. Wang, Y.: Formal cognitive models of data, information, knowledge, and intelligence. Trans. Comput **14**, 770–781 (2015)
42. Wang, Y.: On denotational mathematics foundations for the next generation of computers: cognitive computers for knowledge processing. J. Adv. Math. Appl. **1**(1), 121–133 (2012)
43. Wang, Y.: On cognitive foundations and mathematical theories of knowledge science. Int. J. Cogn. Inform. Nat. Intell. **10**(2), 1–24 (2016)
44. Wang, Y.: On cognitive foundations of creativity and the cognitive process of creation. Int. J. Cogn. Informat. Nat. Intell. **3**(4), 1–18 (2009). https://doi.org/10.1109/COGINF.2008.4639157
45. Wang, Y., et al.: On built-in tests reuse in object-oriented framework design. ACM Comput. Surv. **32**(1es), 7–12 (2000)
46. Wang, Y.: On cognitive models of causal inferences and causation networks. Int. J. Software Sci. Comput. Intell. **3**(1), 50–60 (2011)
47. Wang, Y.: The theoretical framework and cognitive process of learning. In: IEEE International Conference on Cognitive Informatics (ICCI 2007), Lake Tahoe, CA., August, pp. 470–479. IEEE CS Press (2007)
48. Pedrycz, W., Wang, Y., Rudas, I.J., Sun F.: Frontiers in artificial intelligence and autonomous systems. Artif. Intell. Auton. Syst. **1**(1), 1–5 (2023). https://doi.org/10.55092/aias20220001

Addressing Dataset Shift for Trustworthy Deep Learning Diagnostic Ultrasound Decision Support

Calvin Zhu[1,2], Michael D. Noseworthy[1,2,3,5],
and Thomas E. Doyle[1,3,4(✉)]

[1] School of Biomedical Engineering, McMaster University, Hamilton, ON, Canada
nosewor@mcmaster.ca
[2] Imaging Research Centre, St. Joseph's Healthcare Hamilton, Hamilton, ON,
Canada
[3] Department of Electrical and Computer Engineering, McMaster University,
Hamilton, ON, Canada
[4] Vector Institute of Artificial Intelligence, Toronto, ON, Canada
doylet@mcmaster.ca
[5] Department of Radiology, McMaster University, Hamilton, ON, Canada

Abstract. Ultrasound (US) is the most widely used medical imaging modality due to its low cost, portability, real time imaging ability and use of non-ionizing radiation. However, unlike other imaging modalities such as CT or MRI, it is a heavily operator dependent, requiring trained expertise to leverage these benefits.

Recently there has been an explosion of interest in AI across the medical community and many are turning to the growing trend of deep learning (DL) models to assist in diagnosis. However, due to possible differences in training and deployment, model performance suffers which can lead to misdiagnosis and operator hesitancy. This issue is known as dataset shift. Two aims to address dataset shift were proposed. The first was to quantify how US operator skill and hardware affects acquired images. The second was to use this skill quantification method to screen and match data to deep learning models to improve performance.

A CAE Healthcare BLUE phantom with mock lesions was scanned by three operators using three different US systems (Siemens S3000, Clarius L15, and Ultrasonix SonixTouch) producing 39013 images. DL models were trained on a specific set to classify the presence of a simulated tumour and tested with data from differing sets. Principle Component Analysis (PCA) for dimension reduction was applied, then K-Means clustering was used to separate images generated by operator and hardware into clusters. This clustering algorithm was then used to screen incoming images during deployment to best match input to an appropriate DL model which is trained specifically to classify that type of operator or hardware.

Results showed a noticeable difference when models were given data from differing datasets with the largest accuracy drop being 81.26% to 31.26%. Overall, operator differences more significantly affected DL model performance. Clustering models had much higher success separating hardware data compared to operator data. The proposed method

M. Gavrilova et al. (Eds.): *Transactions on Computational Science XL*, LNCS 13850, pp. 110–128, 2023.
https://doi.org/10.1007/978-3-662-67868-8_7

reflects this result with a much higher accuracy across the hardware test set compared to the operator data.

Keywords: Machine Learning · Deep Learning · k-Means Clustering · Ultrasound · Hardware Variance · Operator Variance

1   Introduction

Ultrasound (US) is the most widely used medical imaging modality for its low cost, portability, real time image presentation and use of non-ionizing radiation. The underlying concept is also simple: send sound waves into tissue measure reflections and reconstruct an image based on intensity and travel time. However, unlike other imaging modalities such as CT or MRI, it is a heavily operator dependent modality, requiring trained expertise to leverage any of the benefits. This leads to subjectivity in assessments and possible variability in diagnostics. This can be especially true in emergency scenarios where trained personnel may not be readily available. While errors in emergency ultrasound are due to multiple factors, many include operator or hardware related errors. These include lack of knowledge of technical equipment, inappropriate choice of ultrasound probe, inadequate image optimization, failure of perception, and overestimation of one's skill [11]. Due to the data driven nature of medical imaging, many theorize the growing field of machine learning, specifically deep learning, may add objectivity to the subjective nature of ultrasound imaging and potentially address this operator dependency [1,7,9].

Machine learning is a broad term that encompasses a wide array of techniques. At its core, machine learning uses mathematics to learn from a large amount of representative data of an event or outcome in order to determine or predict the event or outcome using similar data. Many studies have already been done to train a Computer Aided Diagnosis (CAD) tool for ultrasound imaging. One example is Park et al., where they trained deep learning and other machine learning based CAD tools which matched the performance of human radiologists in identifying thyroid nodules in ultrasound images [10].

The operator dependent nature of the imaging modality is a unique hurdle to successful implementation and trust in AI assisted diagnostics. This is largely due to the idea of dataset shift. Dataset shift refers to differences in training data and deployment. Medical AI typically suffers in performance due to changing subjects and demographic differences. Compounding this issue further, ultrasound has high variability across institutions and hardware manufacturers [7] leading to uncertainty that a trained model would be able to fully generalize. Studies are typically done at only one institution with one ultrasound machine [1] suggesting that models may not account for incoming data from differing hardware and operators found at other institutions. The impact of this on model performance is therefore not clearly defined.

In this paper a method is proposed that aims to quantify US operator skill level and hardware differences in order to screen and match data to deep learning models to improve performance. This was done by using unsupervised machine learning to separate images generated by various operators into clusters and computing the distance between these separated clusters. This clustering algorithm can then be used to classify incoming images during deployment to best match the input to an appropriate deep learning model which is training specifically to classify the type of operator or hardware.

2   Background

This section provides an overview of the basic principles of ultrasound, deep learning, and the various algorithms that were implemented.

2.1   Ultrasound

The fundamental principle behind ultrasound imaging is the transmission of ultrasonic waves. These are typically transmitted as pulses and is known as pulsed ultrasound. Pulse repetition frequency (PRF) describes the number of pulses per unit time [4]. Typically, these pulses are 1ms in length and multiple pulses are emitted per second [2]. Since these pulses travel in straight lines, they are often referred to as beams. The direction of US propagation along the beam is called the axial direction, and the direction perpendicular to the axial direction is called lateral.

Clinical ultrasonic waves are high frequency sound waves, typically from 1–20 MHz [4], that are inaudible as the threshold for human hearing is about 20 kHz. The wavelength is inversely proportional to frequency, this means that high frequency waves have correspondingly low wave length and vice versa. Higher frequency waves (10–15 MHz) result in a higher number of waves of compression for a given distance and more accurately differentiates between two structures along the axial direction. Thus, high frequency waves have higher axial resolution [4]. However, higher frequency waves are also prone to higher attenuation and therefore have less ability to penetrate deeper into tissue. This means that higher frequency waves are more suitable for superficial tissue imaging. In contrast, lower frequency waves (2–5 MHz) have lower attenuation and can be used to image deeper structures albeit with lower axial resolution [4].

These waves are transmitted into the subject and reflections back to the source are recorded. When ultrasonic waves travel through tissue, the wave can either reflect, refract, transmit to deeper tissue, or transform into heat. This is due to the fact that sound waves travel at different speeds through differing media. The behaviour of the waves is determined by propagation speed. Propagation speed refers to the speed at which sound can travel through a medium and is typically considered to be 1540m/s in soft tissue [2]. This speed is determined solely by an intrinsic, physical property known as acoustic impedance and is defined as the density multiplied by the velocity of the ultrasound wave propagation in the medium [4]. Air containing tissues such as the lungs have the lowest impedance and dense tissues such as bone have the highest impedance.

Reflections of ultrasound beams form the core of this imaging technique and these reflection beams are commonly referred to as echoes. Reflection occurs at boundaries between two materials provided the acoustic impedance is sufficiently different. Of note is that the only requirement is a difference in acoustic impedance. This means that regardless if the beams are travelling from a higher to lower impedance or a lower to higher impedance medium, an echo will still occur. If the difference between the materials is small, a weak echo is reflected and most of the energy is transmitted deeper into the tissue. If the difference is large, a strong echo is reflected and little energy is transmitted deeper. If the difference is large enough, such as an interface with air, all of the US beam is reflected and no energy is transmitted deeper. Typically in soft tissue, only a small percentage of the beam is reflected. After reconstruction, strong echoes appear as white and weaker echoes are shown as gray.

Refraction occurs when the ultrasound beam hits a surface at angle rather than at 90°C. In this scenario, an echo returns at an angle equal to the incident while the rest of the beam transmits deeper at an angle that deviates from the incident angle determined by Snell's Law 1. Figure 1 shows a visual example of Snell's Law.

$$n_1 sin(\theta_1) = n_2 sin(\theta_2) \tag{1}$$

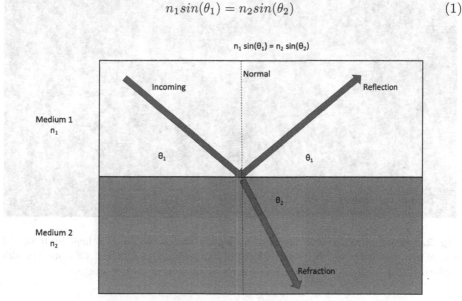

Fig. 1. Visual representation of Snell's law. Snell's law states that an incoming ultrasound wave will deviate at a medium boundary defined by the given equation where n_1 and n_2 are the acoustic impedances of the media, θ_1 is the angle of the incoming wave and θ_2 is the angle of the refracted wave.

Refraction can commonly cause confusion since the reconstruction assumes that the US beams travel in straight lines without deviation. However, this is only in the idealized case and refraction is a useful property for imaging irregularly shaped objects [2]. Scattering occurs when the beam hits an uneven or rough surface and the echoes reflect in various directions. This allows for some of the echo to reach the transducer, allowing for irregular objects to be imaged.

Transmission occurs through uniform tissue or at tissue boundaries of identical acoustic impedance. The wave simply continues into deeper regions until it meets another boundary. However, the wave may lose small amounts of energy as it may transform into heat as the kinetic energy is absorbed by particles.

All of this allows for the reconstruction of an image based on the differing arrival times and amplitude of the echoes. An example image created through ultrasound imaging is seen in Fig. 2.

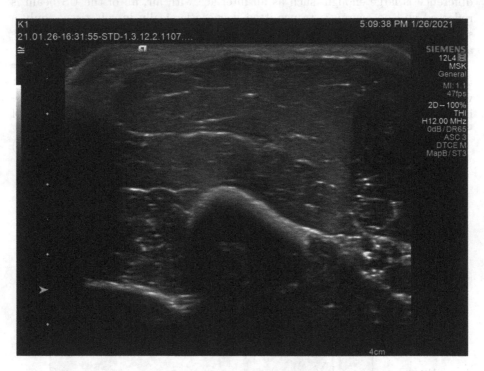

Fig. 2. Sample ultrasound image. Image depicts biceps muscle of a healthy male volunteer scanned with the Siemens S3000 12L4 transducer at 12 MHz. Of note, the skin surface (location of the tranducer) is always at the top of an US image.

2.2 Image Interpretation and Artifacts

There are four types of ultrasound scans commonly used for diagnosis, A-Mode, B-Mode, M-Mode, and Doppler [3]. In A-Mode, a transducer scans a line and plots echoes as a function of depth. In B-Mode, a linear array is used to image a plane through the body which is reconstructed as a 2D image. M-mode refers to motion and is used to capture a range of motion. Doppler ultrasound takes advantage of the Doppler effect to image and quantify flow (e.g. blood flow). Most ultrasound imaging is done in real time B-mode, also referred to as brightness mode. During the reconstruction of an ultrasound image, brighter regions are referred to as hyperechoic and darker regions are referred to as hypoechoic. This corresponds to weaker and stronger echos respectively. Isoechoic regions have echoes equivalent to neighbouring tissue and anechoic regions appear as black regions without echoes.

Artifacts are inappropriate image features that are not true to the object being imaged thus leading to some form of inaccurate representation in image space. Artifacts often arise due to assumptions made during collection and reconstruction. Some of the key assumptions are that sound waves travel strictly in a straight line, that reflections occur from structures along the central axis of the beam, that the intensity of the reflection corresponds to the reflector scattering strength, that the speed of sound in all tissues is exactly 1540 m/s, and that sound waves will travel directly to the reflector and back [2]. In practice these assumptions are made false. Artifacts are helpful sometimes and detrimental other times. It is up to the operator to discern when they occur and how to interpret them (i.e. differentiate them from truth). It is in part why ultrasound imaging is so operator dependent as it is difficult to determine what the image is properly depicting without context.

2.3   Deep Learning

Representation learning is a set of methods that allow a machine to be fed raw data and to automatically discover the trends and representations required to model the data. Deep learning is a form of representation learning with multiple levels. Each level learns more abstract representations of the previous level [6]. Through this method, deep learning aims to discover structure in large datasets typically using a back-propagation algorithm to indicate how internal parameters called weights should be adjusted to best compute the representations in each layer. Each layer is composed of processing units which form a network.

These processing units are commonly referred to as artificial neurons and derive their name from biological neurons. Likewise, combinations of them are named neural networks. Biological neurons produce small electrical signals, called action potentials that trigger the release of chemical signals, called neurotransmitters that in turn affect neighbouring neurons to either hyper or hypopolarize. Similarly, artificial neurons process inputs, and if the inputs reach a designated threshold, they produce an output to other artificial neurons. A neural network is referred to as 'deep' when it has many layers, the exact number of which is typically arbitrary.

2.4   Principle Component Analysis

Conventional machine learning methods, like k-means clustering, typically do not handle raw data very well and as such a feature extractor is required. Principle component analysis (PCA) is a common feature extraction and dimensionality reduction tool used for many machine learning applications. PCA works by selecting a hyperplane closest to the data and projecting the data onto an axis along it. By projecting the data onto a hyperplane that minimizes the mean squared distance between the original data and its projection onto the axis, most of the variance in the data is preserved. The data projected onto this hyperplane is referred to as the first principle component. The data is then projected onto axes on hyperplanes orthogonal to this plane forming additional principle components and preserving the rest of the variance in the data [5].

This whole method is done by singular value decomposition (SVD) described by Eq. 2.

$$X = U\Sigma V^T \tag{2}$$

where X is the training set matrix and V^T contains unit vectors that define all principle components.

2.5   K-Means Clustering

The implementation of K-Means clustering uses a K-Means++ initialization which tends to select centroids further apart. The algorithm is as follows:

1. Take one centroid, c_1, chosen uniformly at random from the dataset
2. Take a new centroid, c_i, choosing an instance x_i with probability

$$\frac{D(x_i)^2}{\Sigma_{j=1}^{m} D(x_j)^2} \tag{3}$$

where $D(x_i)$ is the distance between xi and the closest centroid that was already chosen
3. Repeat for k clusters

Furthermore, to reduce the likelihood of convergence to local minima, the K-Means algorithm is repeated by default 10 times and the result with the lowest inertia is returned. Inertia, in this context, is defined as the mean squared distance between each instance and its closest centroid.

The performance of these clustering models are evaluated using the silhouette score. Silhouette score is a method for cluster analysis that shows which instances are well placed in a cluster and those instances which are somewhere between clusters [12]. Clusters are represented with silhouettes where average silhouette width is a means for evaluating cluster validity and can be used to select an appropriate amount of clusters. Silhouettes are constructed from cluster labels and the distances between instances computed by Eq. 4.

$$s(i) = \frac{b(i) - a(i)}{max(a(i), b(i))} \tag{4}$$

where i is an instance in a cluster, a(i) is the average dissimilarity (distance) of i and all other objects in the same cluster, and b(i) is given by Eq. 5.

$$b(i) = minimum(d_i, C) \tag{5}$$

where we calculate differences between an assigned cluster A and any arbitrary cluster C, i.e. $C \neq A$ and (d_i, C) is the average distance from the instance to all values in another cluster.

The silhouette score is bounded between $-1 \leq s(i) \leq 1$ and scores close to -1 indicate a likely mis-classification, scores close to 0 indicate a decision boundary, and scores close to 1 indicate that the instance is close to the cluster center.

2.6 Dataset Shift

Dataset shift is a concept that analyzes data quality. Dataset shift occurs when unseen testing data experiences events that lead to a change in the distribution of a feature, a combination of features, or class boundaries and as a result, the assumption that the training data and testing data are representative of the same distribution is violated [8].

If a classification problem is defined as a set of X features used to produce a target variable Y, then dataset shift is a situation in which the distribution of X_{train} is not equal to the distribution of X_{test}, therefore leading to an invalid relationship between X and Y.

Dataset shift is a classic problem that can occur in medical AI. Medical AI typically suffers in performance due to changing subjects as populations are not static. Ultrasound introduces many other avenues for dataset shift due to its operator dependency. Ultrasound has high variability across institutions and hardware manufacturers [7]. Furthermore, many studies are typically done at only one institution with one ultrasound machine [1]. Thus, US model performance can vary greatly and there is no certainty that a trained model would fully generalize regardless of the differing operators and hardware.

3 Methods and Materials

This section provides details of the hardware, operators, acquired data, and experimental set-up.

3.1 Hardware

There were three different ultrasound machines deployed:
Siemens S3000 Ultrasound (Siemens Healthcare, Erlangen Germany) using the 12L4 Transducer. This machine saved files in an .avi format with a resolution of 1024×768.
Clarius Ultrasound (Clarius Mobile Health, Vancouver Canada) with the L15 Transducer. This machine also saved files in the .avi format and had options for resolution. A resolution of 1024×768 was selected for consistency.
SonixTouch BK Medical Ultrasonix SonixTouch (BK Medical Ltd., Herlev Denmark) ultrasound (SXTCH3.1-1012.0.11, software Version 6.07) with the L14-5 Transducer was used. This machine saved files as MPEG also with a 1024×768 resolution.

3.2 Phantom

CAE Blue Phantom The primary subject for this work was a soft tissue mimicking phantom from CAE Healthcare. (Sarasota, FL) Specifically, the 'Soft

Tissue Biopsy Ultrasound Training Block Model (SKU NUMBER: BPTM130)' was selected. The phantom contains 16 masses of varying sizes (4–11 mm in diameter) that are hypoechoic or hyperechoic in nature, relative to the gel they are embedded in.

3.3   Operators

There were three skill levels of operators enlisted:

Operator 1 was an engineering student with no formal training in ultrasound operation and minimal medical imaging experience. The purpose of this operator was to closely represent an operator with *minimal technical skill* with ultrasound imaging.

Operator 2 was a medical imaging expert with other modalities, but with no formal training in ultrasound operation. The purpose of this operator was to represent an *intermediate technical skill* level.

Operator 3 was an ultrasound technologist with 3 years of formal training at the time of the study and 8 months of clinical experience. The purpose of this operator was to represent trained personnel with a *high technical skill* level.

3.4   Experimental Set-Up

The data were grouped into various sets in order to facilitate ease of data flow. Tables 1 and 2 show how the data were grouped. Single frames from examples generated before pre-processing from the expert operator showcasing the mock tumour and healthy tissue from each machine is shown in Fig. 3.

Table 1. Data Sets from Operator 3.

Dataset	Train Set	Testing Set
Sonix	4836	1193
Siemens	5462	1572
Clarius	5772	1567
Total	16070	4332

Table 2. Data Sets from Siemens S3000.

Dataset	Train Set	Testing Set
Operator 1	5041	1412
Operator 2	3954	1170
Operator 3	5462	1572
Total	14457	4154

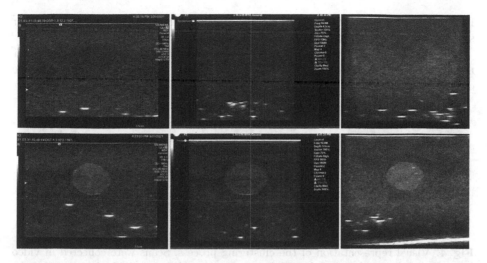

Fig. 3. Single frames collected from the expert operator showcasing the mock tumour and healthy tissue from each machine. From left to right, Siemens S3000, SonixTouch, Clarius.

Models were first trained given data acquired from one machine only. In this first step, the operator was kept consistent to ensure that differences in performance would primarily come from the hardware differences. Data from Operator 3 was used. The expert level operator was selected to minimize errors in scanning and ensure differences were due to hardware differences. Once training was complete, the models were tested on data from the other machines to observe performance.

Next, models were trained given data acquired from one operator only. In this second step, the machine was kept consistent to ensure that differences in performance would primarily come from the operator differences. Data from the Siemens S3000 was used. Once training was complete, the models were tested on data from the other operators to observe performance.

Once base performance was established, the data was then clustered. PCA was first applied to the datasets for dimension reduction and feature extraction. The K-Means algorithm was then ran on the data from the three machines. The operator was kept consistent and data from Operator 3 was used. The number of clusters generated ranged from 2 to 6. Then this was repeated on data from the three operators where the machine was kept consistent and data from the Siemens S3000 was used. The number of clusters generated also ranged from 2 to 6. Figure 4 shows a visual representation of the cluster generating process. The same data in Tables 1 and 2 were given to the clustering algorithm to create clusters.

By using the clustering model as a screening tool, an end user is given more information about the model's decisions by which to trust the results or not.

Furthermore, the models themselves will in theory be more resistant to dataset shift as they are only given data similar to what they were trained on.

To evaluate this, incoming test data was first given to the K-Means cluster model and the Euclidean distance from each cluster center was calculated. The data was then given to the model of the closest center (Fig. 5).

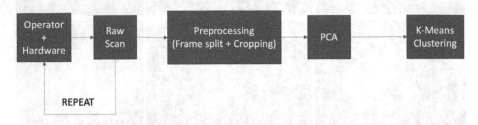

Fig. 4. Visual representation of the clustering process. Scans were collected in video form. During pre-processing the videos were converted into individual frames and the system UI was cropped out. PCA was applied for dimension reduction, which was followed by clustering. This process was repeated with differing combinations of hardware and operators to generate from 2 to 6 clusters.

4   Results

This section provides details of the experimental results collected from initial training and testing, clustering, and the proposed operator and hardware quantification method.

4.1   Initial Training and Test Results

Models were first trained and tested within distribution, i.e. the training and test data were from the same operator and machine. These results were used as a base to compare performance when tested out of distribution, i.e. the test data from a different machine. Table 3 shows the in-distribution results for the hardware models. Table 4 shows the in-distribution results from the operator models.

4.2   Out of Distribution Testing

Each model was tested with the theorized out of distribution test set from different US systems while maintaining the same operator, operator 3. Tables 5, 6, 7, show the test results for the Xception, ResNet50, and VGG19 models respectively. The process was repeated with the hardware models. Tables 8, 9, 10, show the test results for the Xception, ResNet50, and VGG19 models respectively.

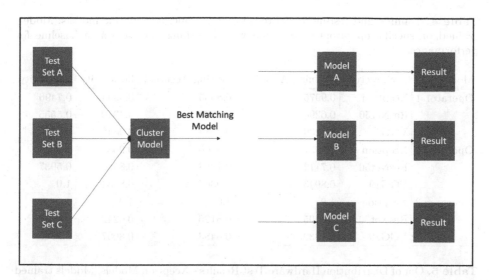

Fig. 5. Proposed method to create more consistent model performance. Test data was first given to a clustering model to determine which model was most appropriate before classification. This addressed dataset shift by better matching samples to the type of data the model was trained with.

Table 3. Training and testing model results for the hardware specific models. Models trained on specific hardware were tested with the matching data sets as a baseline for performance.

Model	Architecture	Training Accuracy	Testing Accuracy	Sensitivity	Specificity
Sonix	Xception	0.9988	1.0	1.0	1.0
	ResNet50	0.7330	0.8878	0.9736	0.6939
	VGG19	0.9376	0.9996	1.0	0.9988
Siemens	Xception	1.00	1.0	1.0	1.0
	ResNet50	0.7995	0.8126	0.6242	0.9410
	VGG19	0.9332	0.8483	0.8097	0.8745
Clarius	Xception	0.9963	0.9948	1.0	0.9894
	ResNet50	0.6831	0.7634	0.5415	1.0
	VGG19	0.9335	0.9616	0.9293	0.9960

Table 4. Training and testing model results for the operator specific models. Models trained on specific operators were tested with matching data sets as a baseline for performance.

Model	Architecture	Training Accuracy	Testing Accuracy	Sensitivity	Specificity
Operator 1	Xception	0.9975	0.8665	0.9984	0.7496
	ResNet50	0.6296	0.4989	0.4350	0.5555
	VGG19	0.9239	0.8388	0.9848	0.7095
Operator 2	Xception	0.9937	0.9700	0.9637	1.0
	ResNet50	0.7112	0.7968	0.8582	0.5037
	VGG19	0.8943	0.9365	0.8834	1.0
Operator 3	Xception	1.0	1.0	1.0	1.0
	ResNet50	0.7995	0.8126	0.6242	0.9410
	VGG19	0.9332	0.8483	0.8097	0.8745

Table 5. Out of Distribution Hardware Test Results - Xception Models. Models trained on specific hardware were tested with the differing data sets to look for differences in performance.

Model	Dataset	Testing Accuracy	Sensitivity	Specificity
Sonix	Sonix	1.0	1.0	1.0
	Siemens	0.9783	1.0	0.9635
	Clarius	0.9475	1.0	0.8916
Siemens	Sonix	0.9659	1.0	0.8890
	Siemens	1.0	1.0	1.0
	Clarius	0.7653	1.0	0.5151
Clarius	Sonix	0.9770	0.9669	1.0
	Siemens	0.7845	1.0	0.6377
	Clarius	0.9948	1.0	0.9894

Table 6. Out of Distribution Hardware Test Results - ResNet50 Models. Models trained on specific hardware were tested with differing data sets to look for differences in performance.

Model	Dataset	Testing Accuracy	Sensitivity	Specificity
Sonix	Sonix	0.8878	0.9736	0.6939
	Siemens	0.8068	0.7437	0.8499
	Clarius	0.5134	0.9467	0.0515
Siemens	Sonix	0.7870	0.7487	0.8738
	Siemens	0.8126	0.6242	0.9410
	Clarius	0.3126	0.5563	0.0528
Clarius	Sonix	0.3752	0.0987	1.0
	Siemens	0.5073	0.5393	0.4855
	Clarius	0.7634	0.5415	1.0

Table 7. Out of Distribution Hardware Test Results - VGG19 Models. Models trained on specific hardware were tested with differing data sets to look for differences in performance.

Model	Dataset	Testing Accuracy	Sensitivity	Specificity
Sonix	Sonix	0.9996	1.0	0.9988
	Siemens	0.8833	0.9323	0.8499
	Clarius	0.5345	1.0	0.0383
Siemens	Sonix	0.8487	1.0	0.5070
	Siemens	0.8483	0.8097	0.8745
	Clarius	0.5319	1.0	0.0330
Clarius	Sonix	0.3860	0.1142	1.0
	Siemens	0.7558	0.4025	0.9967
	Clarius	0.9616	0.9293	0.9960

Table 8. Out of Distribution Operator Test Results - Xception Models. Models trained on specific operators were tested with differing data sets to look for differences in performance.

Model	Dataset	Testing Accuracy	Sensitivity	Specificity
Operator 1	Operator 1	0.8665	0.9984	0.7496
	Operator 2	0.9947	0.9937	1.0
	Operator 3	0.5015	1.0	0.1618
Operator 2	Operator 1	0.9609	0.9516	0.9692
	Operator 2	0.9700	0.9637	1.0
	Operator 3	0.9987	1.0	0.9978
Operator 3	Operator 1	0.8112	0.9592	0.6800
	Operator 2	0.7773	0.8251	0.5488
	Operator 3	1.0	1.0	1.0

Table 9. Out of Distribution Operator Test Results - ResNet50 Models. Models trained on specific operators were tested with differing data sets to look for differences in performance.

Model	Dataset	Testing Accuracy	Sensitivity	Specificity
Operator 1	Operator 1	0.4989	0.4350	0.5555
	Operator 2	0.3919	0.4283	0.2180
	Operator 3	0.6908	0.2374	1.0
Operator 2	Operator 1	0.5344	0.7859	0.3119
	Operator 2	0.7968	0.8582	0.5037
	Operator 3	0.6934	0.4292	0.8735
Operator 3	Operator 1	0.4705	0.9486	0.0468
	Operator 2	0.7656	0.9259	0
	Operator 3	0.8126	0.6264	0.9410

Table 10. Out of Distribution Operator Test Results - VGG19 Models. Models trained on specific operators were tested with differing data sets to look for differences in performance.

Model	Dataset	Testing Accuracy	Sensitivity	Specificity
Operator 1	Operator 1	0.8388	0.9848	0.7095
	Operator 2	0.9425	0.8944	1.0
	Operator 3	0.4149	0.9842	0.0267
Operator 2	Operator 1	0.9041	0.9138	0.8955
	Operator 2	0.9365	0.8834	1.0
	Operator 3	0.8355	0.7783	0.8745
Operator 3	Operator 1	0.5805	1.0	0.2088
	Operator 2	0.5441	1.0	0
	Operator 3	0.8483	0.8097	0.8745

4.3   Clustering

Using PCA the dimensionality of the hardware data was able to be reduced to 175 components while preserving 95% of the variance. In terms of the operator data PCA was used to successfully reduce the dimensionality of that data to 482 components while preserving 95% of the variance.

The K-Means algorithm was run repeatedly to generate n = 2 to n = 6 clusters with data from Operator 3 using all three machines. The expert level operator was selected to minimize errors in scanning. The number of clusters was selected to range from the minimum number of potential expected clusters, 2, to the maximum number of potential expected clusters, 6. n = 2 clusters corresponded to a split between normal and abnormal tissue scans. n = 6 clusters corresponded to a split between 3 machines, each with a split of normal and abnormal tissue scans. Figure 5 shows the results for the expected n = 3 clusters. The mis-classification rate for the n = 3 clusters model was 0.597%. For n = 2 through 6, the average silhouette scores were 0.3446, 0.2683, 0.2638, 0.2885 and 0.2923, respectively. Meanwhile, the average distance between clusters was 45.3892.

The K-Means algorithm was once again ran repeatedly to generate n = 2 to n = 6 clusters but with data from all operators using the Siemens S3000 machine. The Siemens S3000 was selected as it was the more modern machine with the most clinical usage. The number of clusters was selected to range from the minimum number of potential expected clusters, 2, to the maximum number of potential expected clusters, 6. n = 2 clusters corresponded to a split between normal and abnormal tissue scans. n = 6 clusters corresponded to a split between 3 operators, each with a split of normal and abnormal tissue scans. Figure 6 shows the results for the expected n = 3 clusters. The mis-classification rate for the n = 3 clusters model was 71.4%. For n = 2 to 6 clusters, the average silhouette scores were 0.1204, 0.1290, 0.1241, 0.1268 and 0.1381, respectively. Here, the average distance between clusters was 37.0233.

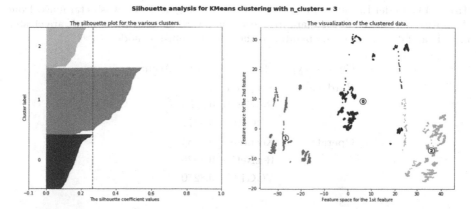

Fig. 6. Silhouette analysis and visual representation of the clusters running K-Means to generate 3 clusters of hardware. On the left is a graphical representation of the silhouette score of each individual sample in each cluster. The red dotted line denotes the average silhouette score. On the right is a visual representation of the clustered data. Note that only the first two principle components are plotted. (Color figure online)

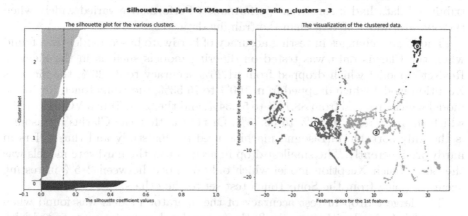

Fig. 7. Silhouette analysis and visual representation of the clusters running K-Means to generate 3 clusters of operators. On the left is a graphical representation of the silhouette score of each individual sample in each cluster. The red dotted line denotes the average silhouette score. On the right is a visual representation of the clustered data. Note that only the first two principle components are plotted. (Color figure online)

4.4 Proposed Method

Each test sample was first passed through the clustering algorithm to identify which cluster it belonged to. From there, the designated deep learning model was used for classification. This process was repeated for each type of architecture, the Xception, ResNet50, and VGG19. Table 11 shows the testing accuracy.

Table 11. Model Results - Proposed Method. Testing accuracy was determined from the entire test set, including data from each hardware dataset in the hardware cluster models and data from each operator in the operator cluster models.

Cluster Type	Model	Testing Accuracy
Hardware	Xception	0.9979
	ResNet50	0.8266
	VGG19	0.9311
Operator	Xception	0.8854
	ResNet50	0.6026
	VGG19	0.8270

5   Discussion

Overall, the models had good in-distribution testing results. The Xception models performed the best with all models having an in distribution testing accuracy of over 86.65%, with the highest being 100%. Models evaluated on 'out of distribution' data had inconsistent performance. Performance varied widely when the operator was different than the training data.

The largest changes in testing accuracy of hardware based models was found when the Clarius data was tested on differing models such as in the Siemens ResNet50 model which dropped from 81.26% accuracy to 31.26%, the Siemens Xception model which dropped from 100% to 76.53%, the SonixTouch ResNet50 model which dropped from 88.78% to 51.34%, and the SonixTouch VGG19 model which dropped from 99.96% to 53.45%. Of note is that the Clarius ultrasound is the only portable ultrasound machine used in this study and thus differs in hardware inherently. The smallest drop in accuracy of the hardware models was the SonixTouch Xception model which only dropped between 2–5% in testing accuracy going from the SonixTouch test set to the others.

The largest drop in test accuracy of the operator models was found when the models trained with operator 3, the expert level operator, were tested with other data. When operator 3 models were given data from other operators, the Xception model dropped in testing accuracy by over 20%, the ResNet50 model dropped in testing accuracy by up to 34%, and the VGG19 model dropped up to 40%. Counter intuitively, some models had increased testing accuracy when evaluated out of distribution which was the case primarily with operator 1, the novice skill level. In the Xception model, the testing accuracy increased from 86.65% to 99.47% when tested with data from operator 2, but sharply decreased to 50.15% when tested with data from operator 3. This result occured again in the VGG19 model where the testing accuracy increased from 83.88% to 94.25% when tested with data from operator 2, but sharply decreased to 41.49% when tested with data from operator 3.

Interestingly, models trained using the data from operator 2, the intermediate skill level, seemed less variable to changes in operators. With operator 2, the

Xception model had performances within 3% performance differences, and the ResNet50 and VGG19 models had performances within 10%. This may be due to an intermediate level operator scanning with elements from both a novice and expert. In this way, the model would theoretically be exposed to tendencies from a wider range of skill levels.

Using the proposed screening method with the operator cluster model the Xception models still performed the best at 88.54% accuracy across the test sets. The VGG19 models had an accuracy of 82.70% and the ResNet50 Models had an accuracy of 60.26%.

The poor performance of the operator models is likely due to the corresponding cluster model not having more training data for each skill level. In theory, the cluster model would act as a proper screening method to ensure that the deep learning model used for classification would see test samples within distribution. However, this is highly dependent on the performance of the cluster model.

6   Conclusion

In this study, operator variability was studied for investigating its effects on deep learning model performance and to determine if this variability could be employed as a screening tool to select the best ML models for the associated skill level and hardware type.

This study proposed using K-Means clustering to screen incoming data and distances between test samples and the generated cluster centres as a means to identify a user skill level. This was primarily motivated by the subjectivity in ultrasonography and the ability for deep learning to provide objectivity.

Overall, it is difficult to conclude whether the proposed methods create a proper means to increase trust for AI assisted ultrasonography and further work may shed more light on this.

Firstly, the operators for this study do not have a diverse range of skill and we believe a better means of quantifying skill level along with more data examples from more operators would improve the results. In review of the results we believe that our medium technical skill level may have too broadly traversed across novice and expert clusters.

Secondly, the screening method(s) could potentially be repeated on patients instead of a phantom such that skill can be better expressed. This is key in separating operator skill as a large part of ultrasonography is using landmarks to identify anatomy and interpreting artifacts. These elements were not present when scanning this commercial phantom as it is consistent and unlikely to produce artifacts.

7   Declarations

The authors have no conflicts of interest to declare that are relevant to the content of this article.

References

1. Akkus, Z., et al.: A survey of deep-learning applications in ultrasound: artificial intelligence-powered ultrasound for improving clinical workflow. J. Am. College of Radiol. **16**(9, Part B), 1318–1328 (2019). https://doi.org/10.1016/j.jacr.2019.06.004, https://www.sciencedirect.com/science/article/pii/S1546144019307112, special Issue: Quality and Data Science
2. Aldrich, J.E.: Basic physics of ultrasound imaging. Critical Care Med. **35**(5) (2007). https://journals.lww.com/ccmjournal/Fulltext/2007/05001/Basic_physics_of_ultrasound_imaging.3.aspx
3. Carovac, A., Smajlovic, F., Junuzovic, D.: Application of ultrasound in medicine. Acta informatica medica: AIM: J. Soc. Med. Inform. Bosnia Herzegovina: casopis Drustva za medicinsku informatiku BiH **19**(3), 168–171 (Sep 2011). https://doi.org/10.5455/aim.2011.19.168-171, https://pubmed.ncbi.nlm.nih.gov/23408755, 23408755[pmid]
4. Chan, V., Perlas, A.: Basics of Ultrasound Imaging, pp. 13–19. Springer, New York, New York, NY (2011). https://doi.org/10.1007/978-1-4419-1681-5_2
5. Géron, A.: Hands-on machine learning with Scikit-Learn and TENSORFLOW: concepts, tools, and techniques to build intelligent systems. O'Reilly Media, Inc. (2019)
6. LeCun, Y., Bengio, Y., Hinton, G.: Deep learning. Nature **521**(7553), 436–444 (2015). https://doi.org/10.1038/nature14539
7. Liu, S., et al.: Deep learning in medical ultrasound analysis: a review. Engineering **5**(2), 261–275 (2019). https://doi.org/10.1016/j.eng.2018.11.020, https://www.sciencedirect.com/science/article/pii/S2095809918301887
8. Moreno-Torres, J.G., Raeder, T., Alaiz-Rodríguez, R., Chawla, N.V., Herrera, F.: A unifying view on dataset shift in classification. Pattern Recogn. **45**(1), 521–530 (2012). https://doi.org/10.1016/j.patcog.2011.06.019, https://www.sciencedirect.com/science/article/pii/S0031320311002901
9. Park, S.H.: Artificial intelligence for ultrasonography: unique opportunities and challenges. Ultrasonography (Seoul, Korea) **40**(1), 3–6 (2021). https://pubmed.ncbi.nlm.nih.gov/33227844, 33227844[pmid]
10. Park, V.Y., et al.: Diagnosis of thyroid nodules: performance of a deep learning convolutional neural network model vs. radiologists. Sci. Rep. **9**(1), 17843 (2019). https://doi.org/10.1038/s41598-019-54434-1
11. Pinto, A., et al.: Sources of error in emergency ultrasonography. Crit. Ultrasound J. **5**(1), S1 (2013). https://doi.org/10.1186/2036-7902-5-S1-S1
12. Rousseeuw, P.J.: Silhouettes: a graphical aid to the interpretation and validation of cluster analysis. J. Comput. Appl. Math. **20**, 53–65 (1987). https://doi.org/10.1016/0377-0427(87)90125-7. https://www.sciencedirect.com/science/article/pii/0377042787901257

Author Index

Printed in the United States
by Baker & Taylor Publisher Services

Printed in the United States
by Baker & Taylor Publisher Services